KV-322-956

012807

Past-into-Present Series

MINING

Hugh Bodey

B.T. BATSFORD LTD, London

First published 1976

ISBN 0 7134 3233 0

Printed by The Anchor Press Ltd, Tiptree, Essex
for the Publishers B. T. Batsford Ltd,
4 Fitzhardinge Street, London W1H 0AH

Contents

Acknowledgment

The Author and Publishers would like to thank the following for their kind permission to reproduce copyright illustrations: the Department of the Environment for figs 2, 3; A. F. Kersting for figs 6, 8, 10, 34; P. T. Bodey for fig. 9; the Science Museum, London, for figs 13, 28 (bottom), 31, 33, 44; D. Bradford Barton Ltd for figs 14, 58; the Mansell Collection for figs 19, 20, 30; the National Coal Board for figs 27, 36, 45, 55; the Ronan Picture Library for figs 28 (top), 29, 40; the British Steel Corporation for figs 38, 48; the Trevithick Society for fig. 39; Radio Times Hulton Picture Library for fig. 43; British Petroleum Ltd for figs 56, 57; and Watts, Blake, Bearne Ltd for figs 21, 54, and 59. Figs 1, 4, 5, 7, 17, 18, 21, 22, 25, 26, 32, 41, 52 and 53 come from the Author's own collection. All the other illustrations appearing in the book are the property of the Publishers.

Chapter 1 The First Miners

1 A diagram illustrating the different mining methods used to obtain minerals from a seam at varying depths. These range from open-cast on the left to shallow pits, bell pits and the beginnings of a shaft mine.

The natural gas and oil lying off the coast of Scotland are the latest in a lengthy list of raw materials that have been extracted from Britain over the centuries. Industrial activity has taken place in Britain for longer than there are written records to describe it, and the basis of this activity has been a plentiful supply of raw materials. Some of these have been deliberately grown or reared, such as the flax that was made into linen or the wool that created the wealth of the country in the Middle Ages. Far greater reserves have been held in the ground in the form of mineral ores, building stone and similar materials. Britain has been fortunate in the wide variety of her deposits, for they have encouraged experiment and led to the development of new industries.

The materials had, of course, to be taken out of the ground before they could be of any use. At first, this was often done by digging the material from the surface of the ground, so creating an open pit. This might properly be called quarrying, or perhaps open-cast mining. The medieval word was 'delving', which applied to any kind of mining. Where the rocks being sought lay deeper than could be reached by surface working, a shaft was cut through the overlying layers and galleries were dug from its base to extract the ore. This method of

2 Grimes Graves as they are now. The 'caves' are the entrances to galleries opening out from the base of the main shaft. Black nodules of flint can be seen in the shaft wall.

working, called shaft mining, presented problems of drainage, ventilation and construction – especially support of the roof. As it was an expensive method, it was only used where the rocks being sought were urgently needed or of some value. Open-cast working and deeper mining have been used for most materials at various times – there are no materials that have been extracted by only one method. For that reason, this book is concerned with all the more common raw materials that have been removed from the ground, whether by normal 'mining' methods or by various forms of open-cast working.

Prehistoric mines

The people of the Stone Age were dependent on stones for all their weapons, tools and domestic implements. Several hard stones were suitable but flint was much the best because it could be split into razor sharp flakes. Flint occurs naturally as rounded nodules embedded in chalk, which could be found lying on the surface in those parts of southern and eastern England where the soil was chalky. Weathering, however, makes rocks break up, so the best nodules were those dug out of the ground. People living in palaeolithic times knew this, and dug shallow pits to obtain fresh nodules. Some of these pits still exist near Brandon on the border of Norfolk and Suffolk. Later, in the New Stone Age, the pits were deepened and new mines sunk. A group of these, known as Grimes Graves, were particularly busily mined for decades before and after 2000 B.C. and are now maintained by the Department of the Environment. One of them has a shaft ten metres deep, with galleries radiating off as far as it was possible to go with the complete lack of ventilation that existed. Roof falls must have been a continuous danger – more than one skeleton has been found of a miner crushed when the gallery caved in. The miners tried to cope with the problems of drainage as best they could. The shaft of a flint mine at Cissbury in Sussex, for instance, was taken a little deeper than the level of the gallery floor, to act as a sump for water. Some of the water would have soaked away through the chalk, the rest was lifted out with a bucket on a rope. The Stone Age communities were careful not to waste anything, and obtained their tools from the butcher – reindeer horns made picks and rakes, and what better for a shovel than the shoulder blade of an ox. Such tools have been found at Grimes Graves and Harrow Hill flint mines near Worthing, and the sides of several mines show the marks of these tools and of the wedges hammered in to split rocks.

The methods of these ancient miners remained in common use for thousands of years. Tools made of bone were slowly replaced by metal heads on wooden handles, but little could be done about drainage and ventilation until the principle of pumps and improved sources of power became known. It was possible to side-step these problems in certain circumstances. In the Bronze Age, for example, tin was mined in parts of Cornwall at some depth (though a horn pick and wooden shovel found at Carnon do not show much advance in the design of tools). This deeper mining was possible because the mines were on cliffs near the sea. When drainage became a problem, narrow tunnels were cut out to the cliff face to drain the water away. Such tunnels are known as adits, and were also used where it was possible to drain into a river.

Bronze is an alloy of tin and copper, and was widely used by the Mediterranean civilizations from about 2500 B.C. At first tin was obtained from the silt in streams draining tin-bearing areas; later the ore deposits in Saxony and Britain were mined and smelted. It is not possible to say when tin mining began in Britain, but Cornish tin was widely famed for its quality by 1000 B.C. and it must have taken a considerable time for production to have built up to such a level as to provide a surplus for export. A smelting furnace was set up on St Michael's Mount, where the tin was cast into ingots to be loaded on board ship in the harbour. Copper ores can be found close to tin in several parts of Cornwall, and it is likely that they were also mined in the Bronze Age, though

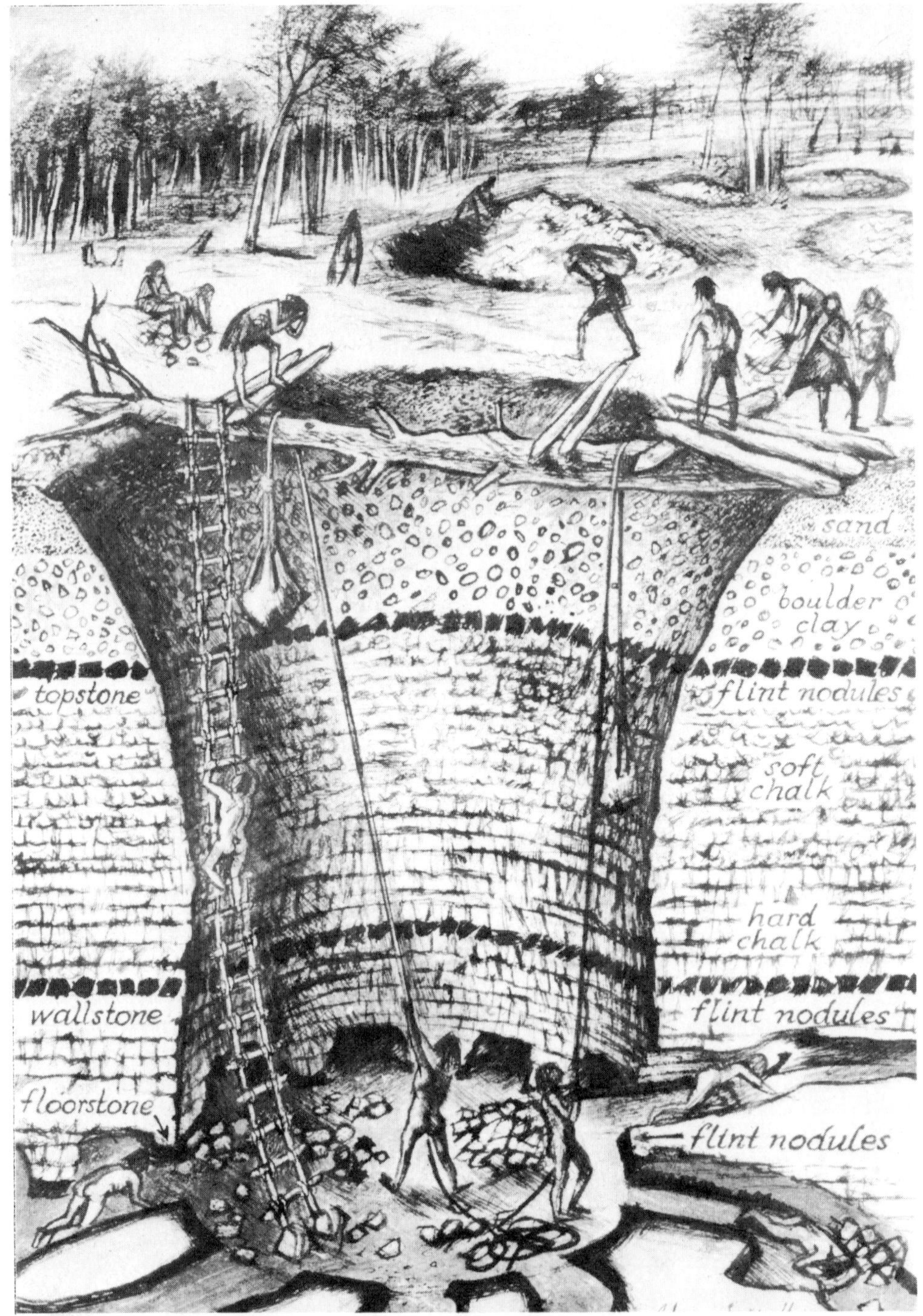

3 An artist's reconstruction of miners at work in Grimes Graves. Notice their way of removing the waste rock, the positions in which the miners must work and the fact that there are as many surface workers needed as miners.

4 A heavy stone measure brought by Phoenician traders to Cornwall when they came to buy tin. This one was lost overboard.

copper was not exported. The Irish were mining copper from around Avoca in the Wicklow Mountains as early as 2000 B.C. They used gold from rivers in that area to buy tin from Cornwall, and alloyed that with the copper to make bronze. The main copper workings in England were at Alderley Edge in Cheshire, and smaller quantities were dug from open-cast trenches in Anglesey, the Lake District and south-west Scotland. Like the Cornish tin, most of this ore was crushed and smelted near the pits from which it was dug, to reduce the cost and difficulty of transport.

Iron Age

Some time after 1000 B.C. miners began to dig lead from the ground. It outcropped on the surface in the Mendips and to a lesser extent in Derbyshire, Flintshire and the Yorkshire dales. It was mostly mined from shallow pits, and was smelted on the spot because of its great weight. Most British lead also contained some silver, and it seems that some of this was separated out. Another precious metal, gold, was recovered from streams in central Wales, though whether any form of mining was undertaken is uncertain. Information is rather more plentiful about iron mining, which began in Britain about 500 B.C. Iron ore outcropped on the surface in many parts of Britain, and was exploited by successive waves of immigrants variously called Celts or Iron Age people. The latter name indicates that these people were the first in Britain to know the value of iron as a strong metal, and had the necessary knowledge and skill to build furnaces that could achieve the high temperatures needed to smelt the ore. The smelting was done with charcoal, which was readily available since Britain was heavily forested then, but it was not an easy process and the Celts could not afford to make everything out of iron. Indeed, it was treated as a precious metal and was used to make currency bars as well as weapons. Copper, bronze and lead continued to be the metals available for ordinary use, though metals could never be produced cheaply enough to be as commonplace as they are in the twentieth century. Tin was also mined by the Celts – they built Chun Castle near St Just in Cornwall, to protect a group of 11 huts in which tin was smelted, and there were other forts like it in the county.

Roman Conquest

All this mining and preparation of metals in Britain attracted the attention of the Romans, who were always on the look-out for new sources of supply. Trade between several British tribes and merchants of the Roman Empire increased slowly from about 100 B.C., but this did not give the Romans sufficient control over output. For this and other reasons the Romans decided to conquer Britain in A.D. 43. All minerals were officially the property of the Empire from that moment, though troops spent many years making imperial control a reality. Their first target was the rich Mendip lead mines, since these would be the best source of silver for currency. They were taken over by A.D. 49. Houses were quickly built at Charterhouse for imperial civil servants who were appointed to organize and increase production, using, as the workforce, an endless supply of expendable slaves drawn from the many prisoners of war and convicts. The smelted lead was cast into ingots or pigs weighing about 45 kilos, and sent to the nearest port, which was Southampton. Customs officials there either loaded the pigs on board ships en route for Rome or diverted them for use around Britain. The same pattern of official supervision was established in each new area as it was conquered: Flintshire by 74, Nidderdale by 81, and later the other Yorkshire dales, Derbyshire and Shropshire.

The purpose behind having resident officials in each mining area was to increase production. It was possible to do this for a limited period simply by drafting in large numbers of 'miners' and working them to death. The key to continuous production, however, lay in prospecting for new veins (or lodes) of ore, so that there would always be known reserves to mine. Roman engineers had considerable experience of mining in other parts of the Empire to guide them, and were able to keep output expanding throughout the first century A.D. This was not easy, for most British ores were in thin veins, which had often undergone

5 The shaded areas on the map show where there was extensive mining or quarrying during the Roman occupation. Single mines, or a small group, were worked at the other sites marked, and the minerals were processed nearby.

INDUSTRIAL SITES IN ROMAN BRITAIN

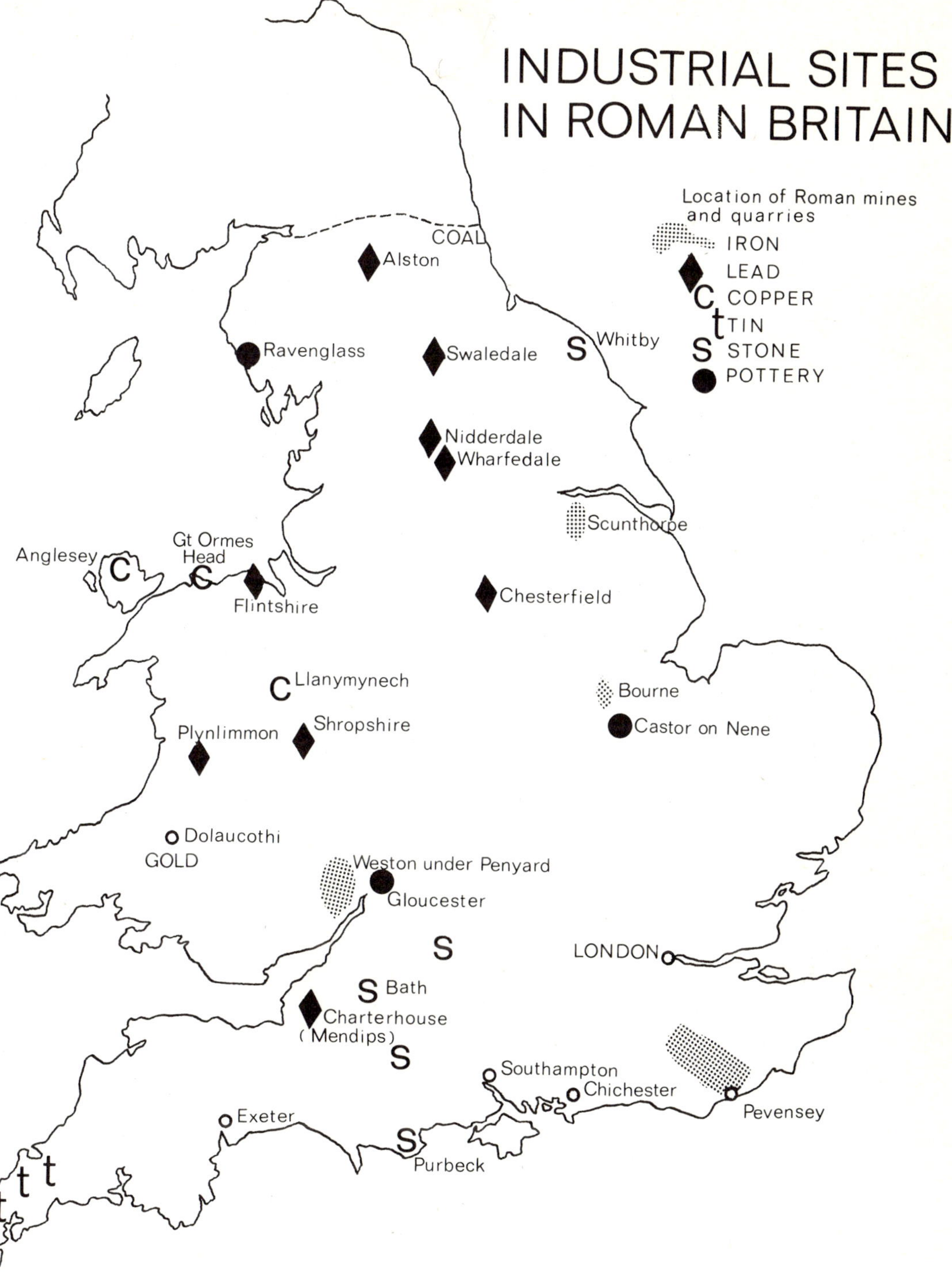

considerable faulting, making them hard to follow. Native miners were used to such problems, so when sufficient tribes seemed to be under control the Romans handed the mines back to British miners. The Yorkshire mines were relinquished in 154, while a group of miners calling themselves the Lutudarum company took over in Derbyshire. The Romans took some of the lead they needed as taxation, and bought the rest. In addition, the legions obtained lead for military purposes from their own mines where this was possible, as on Alston Moor in Northumberland and the slopes of Plynlimmon in Cardigan. While these legionary mines were only worked to supply immediate needs, lead was mined from the principal areas throughout the four centuries that the Romans governed Britain. Most of the lead was dug from open-cast pits and shallow workings, supplemented by bell pits when the surface deposits had been exhausted.

Lead was mined with such zeal and efficiency because of its valuable silver content; the Romans were in less of a hurry to reorganize mining of other minerals. They had little need for tin because of the rich Rio Tinto mines in Spain, and so they did not seriously attempt to colonize Cornwall. Later, when in the third century the Spanish mines became too costly to work, the Romans seemed content to buy ingots of tin from Cornish metal merchants. Cornish smelting was very efficient – a dish made about A.D. 350 was more than 99 per cent pure tin. Silver-bearing (argentiferous) copper had been mined near Callington in the first century B.C. but was abandoned about 100 years later – Roman engineers who visited Cornwall about A.D. 70 found nothing but tin, and developed copper mining in what seemed more likely areas.

There were several of these, the richest being the Parys Mountain mines on the island of Anglesey. This was also the headquarters of the Druid religion, and copper was paid over as tribute to the Romans to avoid invasion. Somewhat less valuable deposits were mined under direct Roman control, similar to the Mendip lead mines. Along the coast from Anglesey, adits were dug into Great Orme's Head and side galleries were cut from these to follow the lodes. By contrast, a large cave was the basis of the mine at Llanymynech, which had been excavated by years of face-working. The slave and convict miners slept in it, while mining was again extended by gallery workings. Copper smelting, which required greater skill than lead, continued to be done by free workers, without close official supervision.

The last of the non-ferrous (other than iron) metals mined at this time was gold, and this was radically developed by Roman civil engineers. Some gold had been obtained by panning in Carmarthenshire rivers for years. The Romans were determined to increase production, particularly from the hillside at Dolau Cothi, near Llandovery. Undercutting was used in the second and third centuries to bring down the rock, and a sizeable quarry was created. Where the overburden of valueless rock was too great to allow open-cast working, adits were cut into the quarry face, and the rock was loosened by fire-setting and the use of hammers and wedges. Water played a large part in the development of Dolau Cothi. Aqueducts 6½ and 11 kilometres long were built to supply water to the site, where it was used to wash the valueless rocks away from the gold-bearing quartz, to carry quarried stone further down the hill and, above all, to separate the heavy grains of gold from lighter rock fragments. It is possible that this was done by washing the metals in fast-running water, and passing it over sheep fleeces at such a speed that the gold sank into the fibres while the other particles were swept on. However the gold was recovered, it is clear that no expense was spared to increase output from the site, though at the cost of the lives of captured tribesmen.

Lead and silver, tin, copper and gold were the metals that most interested the Romans. We know about these deposits and the methods of mining them because of the Roman habit of keeping records and constructing permanent buildings near the sites. Less is known about iron mining at this time since it was of slightly less value and the Romans had sufficient supplies elsewhere. As a result most of it was worked by British miners and smelters who left few records. The Romans took over the mines in the Weald of Sussex soon after the invasion of A.D. 43, and finally captured those in the Forest of Dean between 74 and 76. These remained their principal mining sites, with iron-working concentrated at Chichester and Weston-under-Penyard, but mines were later opened up in Northamptonshire, Lincolnshire and a few other places. Iron was solely a practical metal for the Romans – it was produced for spearheads, pickaxes and farming tools, and had no value in itself. This tended to debase the value of iron currency bars among the British tribes, who turned to using copper and silver instead.

Stone

The British had ignored the rich variety of building stone in Britain but the Romans, long accustomed to building in stone, made the most of what they found. A quarry was opened near Bath in A.D. 61 for the creamy limestone often called Bath stone. This could be cut to a flat surface or carved into mouldings, and was used for nearly all the public buildings in the area. In addition, it was ferried across the Severn for a temple at Lydney, taken to London for many other buildings and even further afield than that. A still finer stone was quarried on the Isle of Purbeck in Dorset. Though not as fine as the marbles used in Italy (some of which were imported) it made an acceptable substitute in many villas and public buildings. Stone from both these quarries could easily be put on board ship and taken to any building sites which were near water. Inland areas were deprived of such stones, in any quantity at least, and so developed the local stone of the area – limestone in the Cotswolds and Northamptonshire, flint in East Anglia, sandstone in the Midlands and so forth. Perhaps the Romans' most lasting contribution to Britain was the opening up of

such a variety of quarries, for this paved the way for the rich range of building materials and therefore styles of architecture which characterized Britain until the modern age of town planning and concrete.

In addition to building stone, gravels were quarried for surfacing the Romans' 8,000 kilometres of British roads, as well as bigger stones for their foundations and kerbs, clay for pottery, bricks and tiles, and sand for the small glass industry in Yorkshire. Some coal was also dug up, though not with much enthusiasm. There was no real shortage of timber for fuel in a country so densely wooded as Britain, and it would seem that coal was only taken if it lay on the surface and could be used nearby. Some mined in Northumberland, for example, was used to warm the troops guarding Hadrian's Wall. In South Wales coal ran a central heating system in Caerwent, and it kept a permanent flame burning in a temple in Bath. However, the low value attached to coal is clear when we learn that empty grain ships were ballasted with it in the Tyne, and that most of it was thrown into the sea when the ships arrived back in the Fens. Few people could have foreseen the importance coal would acquire in later centuries.

6 *Opposite* This picture of Bath, Avon, illustrates a variety of the uses of stone over many centuries. In the foreground are the baths built by the Romans, further away the medieval abbey, and buildings dating from the eighteenth and nineteenth centuries. Stone has been used for basic construction, decoration, statues and roofing.

7 *Below* The foundations of a Roman road at Blackpool Bridge in the Forest of Dean. The kerb stones and other stones have been deliberately placed to ensure stability, and were originally covered by a surface of crushed iron slag.

Chapter 2 Mining in the Middle Ages

The last of the Roman garrisons left Britain in A.D. 410, and six centuries of uncertain turmoil followed. Warriors from Saxony had made a number of raids as early as the fourth century, but had been repulsed by Roman troops and local residents. The raiders were more successful once the Roman forces had gone and, spurred on by overpopulation at home, followed up successful raids with the creation of colonies. These quickly spread and soon covered most of southern England from the Bristol Channel to Kent. During the fifth and sixth centuries boatloads of Angles reached the shores of Britain and began to settle in Norfolk and Suffolk (East Anglia): by 800 Middle Anglia was a separate kingdom in the Midlands. At the very end of the eighth century the Viking raids began with a few isolated attacks but soon became a serious menace. By 865 the Vikings (from Denmark) had established an area known as the Danelaw, covering the better farming land in a broad strip of eastern Britain. Here they settled down as farmers, and started levying taxes on neighbouring kingdoms. Norwegian Vikings, meanwhile, had colonized the Isle of Man, Galloway and many other scattered districts, mostly in the underpopulated north-west.

These raids, invasions and conquests lasted for more than two centuries and brought havoc and destruction. Some of the resident Celts moved to the Roman towns in search of work, but most preferred to remain in their hill-top villages which they could defend from outside attack. In eastern England, however, the land was flat, and there were few defensive sites. As a result many Celts were forced to retreat. Some stayed behind, intermarried with the new arrivals and became subjects in the new kingdoms. Those Celts who would rather die than submit to any invader were slowly forced into the more mountainous, or at any rate poorer, farming areas – Cornwall, Wales, Cumbria, Ireland and southern Scotland.

Dark Ages

It is hard to say with any great certainty what happened to the mining industries during the six centuries after 410. In some areas the Romans had ceased to manage the mines in the second century, and Roman management had been withdrawn everywhere by A.D. 300. In theory, the produce of the mines was still imperial property, but increasingly in practice the Romans merely levied a

8 The Saxon church at Bradford-on-Avon, Wiltshire. Notice how the blocks of stone have been cut to fit together, the window and door arches, and the stone 'slates'.

tax on output. The departure of the Romans seems to have caused a reduction in the amount of metals mined, partly because there were no longer the taxes to pay and also because the number of merchants buying goods for sale in other parts of the Empire declined. Stone quarrying abruptly ceased – no one bothered to maintain the roads, and only a handful of people were sufficiently impressed with Roman building styles to attempt to maintain them. It seems equally unlikely that any coal was dug up throughout the Dark Ages, save possibly in the Tyne and Wear valleys where it lay on the surface.

However, a few exceptions stand out from this general picture of decline in mining activity. The Cornish, who were almost untouched by all the conquests, continued to mine tin, and to sell the smelted ingots to many parts of Europe and to Mediterranean countries. It is highly likely that Celtic miners in Wales and Cumbria continued to mine small quantities of copper, and also lead (and therefore silver) and gold. A proper system of coinage slowly emerged during these centuries, and most of the reserves of metal for this lay under Celtic control.

The almost constant fighting also demanded large supplies of iron. Iron ore could be mined in many parts of Britain, such as the Midlands and South Wales, as well as from the large deposits in Gloucestershire and Sussex. Archaeological evidence indicates the use of iron for swords, daggers, shields and helmets, and also for a few highly prized tools. The invaders needed weapons and tools as much as the defenders, and it would seem that they were at least as skilled in the business of prospecting and quarrying for mineral ores, smelting them and manufacturing goods from the metals. The invaders probably continued to work mines in the areas they took over. The laws of the Derbyshire lead miners, for example, include a number of Saxon words and titles for their offices – the Barmaster is derived from Bergmeister, for example. Mines in many areas must have changed hands between 500 B.C. and A.D. 1000. Though the upheaval was considerable in the country as a whole, it is unlikely to have troubled any particular mine very often. The rich quality of manufactured goods that survive suggest that craftsmen had sufficient peace and quiet in which to develop their skills.

Stone

Two industries, pottery and building, suggest some measure of growth during the Dark Ages. Nearly all settlements used pottery bowls and jars for cooking and storing food. This required supplies of suitable clay, which had to be dug up. The only alternatives were glass containers – sand was quarried in Kent for glass-making and later near Glastonbury, but the quality of the glass was poor.

It seems that the invaders had rather grander ideas than the Celts on the matter of building, derived from what they had seen in other places. So, when St Benedict wanted to build a new church in about 650, he sent to Gaul for some masons and glaziers to supervise the work. Soon after, in 675, St Aldhelm began replacing the wooden abbeys of Malmesbury and Sherborne with buildings of Cotswold limestone. In the north of England, some Saxon villages used the local gritstone in their public buildings. Their tools were not very suitable for so hard a stone, and the work needed more patience than skill. Surviving buildings, such as the little church at Bradford-on-Avon, add to the evidence that the Saxons were able to find suitable stone, and quarry and shape it, in order to build for the future. Some of their buildings were also for defence. Many churches in East Anglia had tall, round towers which were used as lookout posts to give notice of Viking raids. However, the church towers were no more successful against the Vikings than the forts had been against the Saxons. Many of the towers were built of flint, which was almost the only hard building stone in the area. Prehistoric mines like Grimes Graves as well as new mines were worked for flint. The Vikings also liked to work in stone. The complex carving they used on wayside crosses and coffin lids indicates that they were using hammers and chisels in their work, and they may also have sawn blocks of stone in the quarry.

9 The Viking settlers quarried large slabs of stone for coffins (as here), wayside crosses and other permanent memorials. Their decorative carving was very distinctive.

The Norman Conquest

The last in this series of invasions of Britain was the Norman Conquest of 1066. The Normans quite quickly subdued the Saxon and Danish kingdoms, though they met with some stubborn opposition from the Celts in the hillier areas. Several centuries were to pass before the whole country acknowledged a Norman as king. The Normans were firm believers in the importance of keeping records; they taxed their new subjects heavily to pay civil servants to do this work, and many detailed records still exist which help to sweep away the haziness of the Dark Ages. In addition, a larger number of the products of the Norman era have survived.

This is particularly true of castles, abbeys and churches, which consumed the output of the most active mining industry in the Middle Ages – stone quarrying. The Normans came to England with some advanced ideas on the construction of large buildings, and these were developed in subsequent centuries to produce some of the most imaginative and adventurous buildings in Britain.

A few of the quarries the Normans used were already being worked when they arrived, but most they sought out and developed themselves. The officials who compiled the Domesday Book (1086) noted that limestone was being quarried in the Windrush valley in Oxfordshire, for instance, and the quarries were able to supply large quantities for the building of Windsor Castle between 1363 and 1368. Monasteries frequently opened up their own quarries – those at Barnack, near Peterborough, were among the largest in England. Stone was sent from them down the River Welland, Carr Dyke (a Roman drainage canal) and across the undrained Fens to several major building projects that could be reached by water. The Bath stone that the Romans had used was taken by water to Winchester for the royal palace in 1221. Sandstone was another stone popular with medieval masons; it was quarried near Bridgnorth in the 1170s for Worcester Cathedral and adit-mined at Beer in Devon in the 1360s for work in Westminster and Rochester. A much harder stone from Portland was used in the construction of Exeter Cathedral and Westminster Abbey, but it was so hard to work that it was only used for the most important buildings. Even harder was the granite of Devon, Cornwall and Leicestershire, which was only used for a few parish churches and castles where no other stone was available.

The Normans preferred the smooth surfaces that could be obtained by sawing stones – their saws could handle limestone and sandstone but Portland stone and granite were beyond their tools. Other stones were quarried to make the roofs of these new buildings. The prior of Winchcombe in Gloucestershire used

10 The west front of Exeter cathedral, built from stone from the mines at Beer. Some stones have stood up well to centuries of weathering, others have crumbled away and the long job of replacement has been started.

11 A slater splitting blocks of sandstone at Collyweston to make roofing slates. He is helped by the natural cleavage of the rock, as can be seen, and his tools and the job itself have not changed since Roman times.

limestone slates in 1221, which were expensive and almost unheard of locally, despite the fact that Roman masons had used them nearby at Chedworth 1,000 years before. Limestone and sandstone slates became more common in the fourteenth century, especially in towns, where thatch was a fire hazard. True slate was quarried for local use in Wales and Cumbria in the thirteenth century, and more widely in Devon and Cornwall. Not only was it used to roof and face the walls of buildings there, but considerable quantities were sent to other ports by sea – 800,000 slates were dispatched to Winchester for the king's buildings in the twelfth century.

Other stones were mined on a smaller scale. Flint was taken from the ground to build the last East Anglian church towers in the twelfth century, and was used long after for public buildings (like Norwich Guildhall) and some large private houses. In the same area a hard kind of chalk called clunch was mined for Ely Cathedral, some of the Cambridge colleges and other places. Considerable quantities were mined from bell pits, locally called dene-holes, in Essex and Kent for spreading on the fields as a fertilizer. More specialized was the marble

12 Slates have to be trimmed to shape after they have been split, and a hole cut so that each one can be nailed to a wooden batten in the roof.

produced in royal quarries on the Isle of Purbeck in Dorset. This was mostly used to ornament cathedrals and gave rise to a lively industry in the thirteenth century; local craftsmen cut and polished the stone at the quarry and sent the completed work to many sites. Fountains Abbey was decorated with local black marble from Nidderdale in the thirteenth century, and many northern churches were adorned with Frosterley marble which could be sent down the River Wear. The Trent was used to transport blocks of alabaster to Nottingham from quarries and mines which opened about 1400 in Staffordshire and Derbyshire. It was mostly used for memorial statues.

A newcomer among quarried rocks, clay, took a long time to find favour. It was used in the twelfth century to make roof tiles, particularly for town houses, to reduce the amount of inflammable thatch in use. Writing about London at the end of the century, William Fitzstephen noted:

> As for prevention of casualties by fire, the houses in this city being then built all of timber, and covered with thatch of straw or reed, it was long

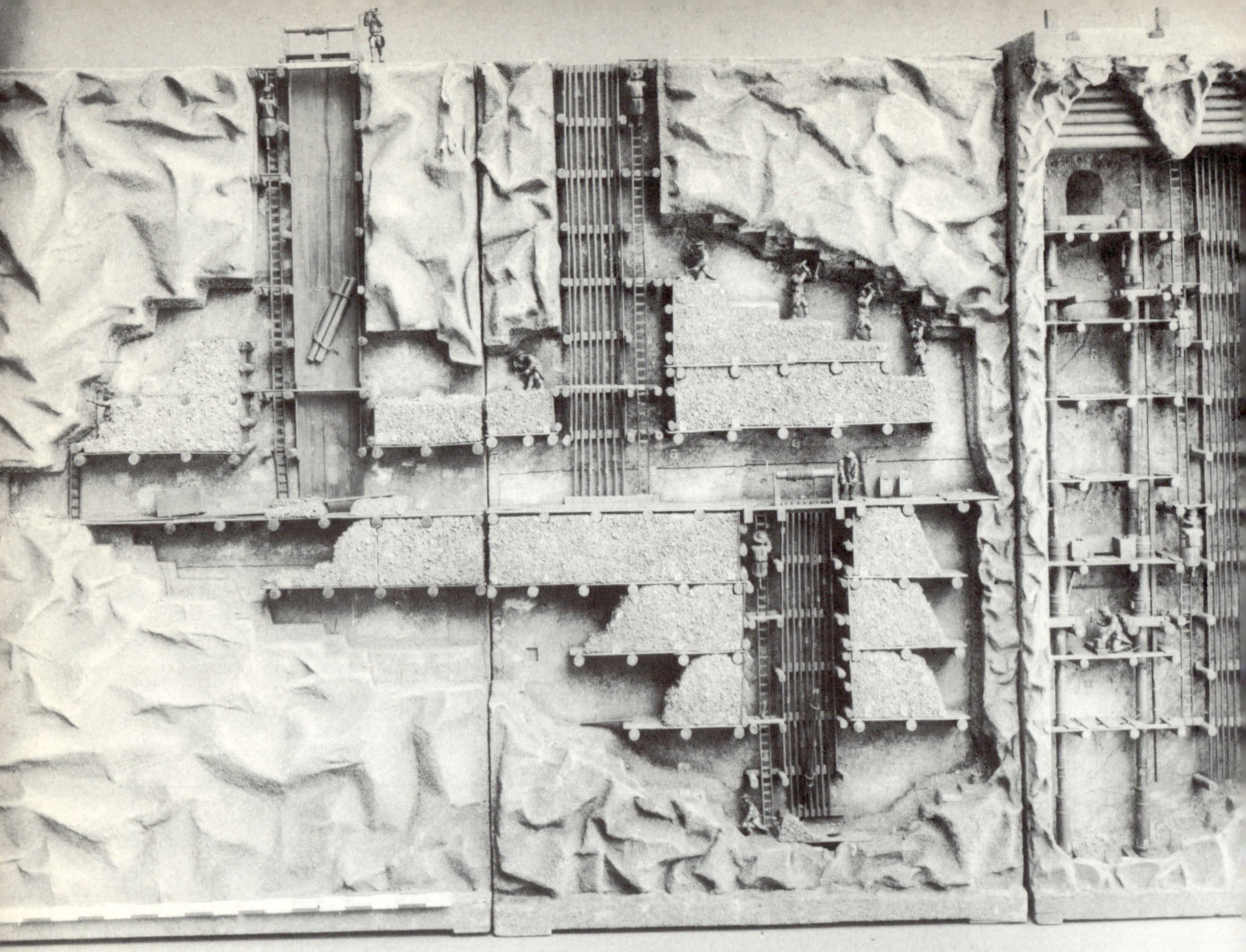

13 A model of the layout of a Saxon mine. The miners in the upper part of the picture are overhand stoping – cutting at the roof and letting the waste fall on to platforms to raise the floor level. The miner below is digging the floor – underhand stoping. The mine is drained by a series of hand pumps, emptying into troughs at each stage; the top trough drains away down an adit. Only the drainage shaft has been sunk to the full depth of the mine, while the working shafts have been dug to follow the direction of the lode.

> since thought good policy . . . to provide . . . in . . . 1189 . . . that all men in this city should build their houses of stone up to a certain height, and to cover them with slate or baked tile; since which time . . . there hath not happened the like often consuming fires in this city as afore.

Tiles were being made in 14 counties in east and south-east England by the end of the thirteenth century, and in Somerset, and the demand for clay continued to grow. Bricks began to be made from clay at about the same time. Flemish immigrants used Colchester clays to make bricks for Little Wenham Hall in Suffolk in 1275, and a number of brick buildings rose around Hull in the fourteenth century. When Henry VI founded Eton College in 1440, a field was bought at Slough and a kiln set up which made 2.5 million bricks between 1442 and 1452. The age of brick building lay mostly in the future though, for there were plentiful supplies of stone and this was still preferred as a building material. While there were a few large quarries like Barnack, the bulk of the stone used at this time came from small, shallow quarries, some of which were worked to build a single house and then abandoned.

Metals

The mining of all non-building materials was on a much smaller scale. Metals were needed for tools and in the home, and lead was used to roof some of the larger buildings; but the increase in the amounts produced was small, just keeping pace with the slow rise in the population. Some tin was still mined in Cornwall, and a little copper in Cumberland and Alderley Edge, until reserves of the fuel needed for smelting were exhausted in both areas. Copper was then imported from Germany and Sweden to supplement home production.

Lead mining was rather more important because of its more varied uses. Monks from the Benedictine abbey at Repton ran mines at Wirksworth in the ninth century, and the mines in that part of Derbyshire provided work for seven smelting furnaces in 1086, supplying 250 pigs of lead for the roof of Westminster Abbey. The area remained the principal lead mining area in the Middle Ages, and local customs were written down in 1288 as a guide to the way the industry was to be run. Lead was also mined in Flintshire, and *Laws and Customs of the Mine,* written in 1352, became the rule book of the miners there. The Mendips continued to be an important lead producing area, though

14 The Panty Oon stone, a large block of gritstone still to be seen on Greenhow Moor in Yorkshire. It was hollowed out in the thirteenth or fourteenth century so that lead ore could be crushed in it, probably by men lifting and dropping iron-shod poles on the ore. Lead mining was organized in the area at that time by the monks of Fountains Abbey – monasteries often led in developing mine sites in the Middle Ages.

output figures do not exist for that or any other district. Lead was vigorously mined in some areas for the silver it contained, especially in times of war. Three hundred Derbyshire miners were taken to Devon on the instructions of Edward I to mine argentiferous lead at Combe Martin, and mines nearby at Bere Alston produced silver valued at £1,773 in 1305 and £180-worth of lead. The feverish activity declined after 1340, though the mines continued to produce some metal until the end of the Middle Ages.

Iron mining also increased, for weapons, general use and to make cards, combs and other tools for the woollen industry. The medieval kings encouraged mining on their estates – Henry III collected rents from his Forest of Dean mines in the 1250s – and so did the monasteries. The Cistercians were particularly skilled at smelting iron, a skill based on knowledge gained in France. The monks of Kirkstead in Lincolnshire obtained a charter to mine ore and gather wood (for charcoal) at Kimberworth in 1160, Fountains Abbey had mines and forges outside Huddersfield, and Tintern Abbey had others in the Forest of Dean. The industry was scattered throughout Britain, but total output did not exceed the country's needs and its cost prevented it being an everyday material.

Coal was almost ignored until the thirteenth century, and total output seldom exceeded 15,000 tonnes a year before 1500. Most of that was dug from adits and bell pits in the Tyne valley and the area to the south, much of which belonged to the Bishop of Durham who was keen to develop his estates. Some of the coal was used locally, and the rest went by ship to France, Flanders and London – Sea Coal Lane existed in 1228. A little mining was done in Lancashire and in those other areas where it lay on or near the surface, but few people could be bothered with it when wood was so plentiful. However, by 1500 supplies of wood were beginning to decline, as it was consumed in industry and to build ships and houses.

15 Abandoned twelfth-century iron pits at Bentley Grange in Yorkshire. The spoil tips have now subsided and are about two metres high. The bushes grow in the hollows which were once the deeper, open pits.

Chapter 3
The Start of Industrialization

The sixteenth century was a time of quite rapid industrial expansion. The woollen industry was the most profitable, employing people in every British county. Exports of woollen cloth increased steadily under the control of the Merchant Adventurers' Company. The growth of the industry required a corresponding growth in the supply of coal for heating washing water and dye vats, of iron and non-ferrous metals to make cards, combs and other tools, and of building materials. Wool cloth production was not the only industry to expand, and growth in all industrial sectors necessarily involved an expansion in mining. Some of the growth was based on new technological methods, many of them brought into the country by immigrants. The introduction of large blast furnaces like those used in Germany, for instance, made it possible for much larger castings to be made and so led to a demand for more iron ore to be mined. New ways of making brass wire, with the aid of water power, led directly to a search for new copper mining sites. The development in industry was encouraged by the closure of the monasteries in the 1530s, which left many men who were used to industrial processes looking for work, and made available a number of sites and buildings that were well suited to industrial working.

Stones

The dissolution of the monasteries by Henry VIII marked the end of the period of building large churches and castles. The long period of civil war (the Wars of the Roses) had ended in 1485 with the defeat of Richard III, and with the return to stable government under Henry VII people turned their energies to building more comfortable country houses. Large windows replaced arrow slits, and massive curtain walls became mere boundary walls. Although stone was no longer needed to build castles and abbeys, many new quarries were opened up in the sixteenth century to supply stone for the new-style homes. The most popular were still sandstone and limestone, which could be cut to flat surfaces without much difficulty. The same stones were cut and split into slates too, and quarrying for real slate was extended in Cornwall and Cumbria.

Other stones were also quarried. Alabaster was used for memorial effigies in some parish churches, for example, while Henry VIII had supplies of stone cannon-shot cut in quarries at Maidstone. As always, the local stone was dug in every area for ordinary buildings, and even for some others, because the cost of transporting any building materials could only be kept within reasonable limits if water transport was at hand. Communities remote from rivers had to make the best of what lay beneath their feet.

In much of southern and eastern England this was clay. The use of bricks to build houses increased steadily in the sixteenth century, and towns as large as Southampton had their own official brick-makers. A few houses were made entirely of brick, but it was more usual to combine it with other materials. There was still sufficient timber in Elizabeth I's day (1558-1603) for merchants to build their houses on a stout timber frame. Bricks were used to fill in the gaps (nogging), in place of the old-fashioned wattle and daub. As timber became more scarce houses were built entirely of brick, sometimes faced with stone. The widespread use of tiles also added to the demand for quarried clay.

Iron

Supplies of iron ore continued to come from the Weald of Sussex and the Forest of Dean. There was no shortage of ore in either area, but the decline of the forests began to cause concern towards the end of the sixteenth century. The Weald concentrated on making the heavy cannon and other iron weapons needed by the government (the Spanish Armada in 1588 was only the beginning of 15 years of war), and supplies of charcoal were withheld from many ordinary people so as to keep enough in reserve for the iron furnaces. The Forest of Dean was also the principal source of timber for the navy, which meant more shortages for the iron smelters; and the forest warders made worried reports about the shrinking acreage of good timber.

Iron ore had long been dug in other places, and the difficulties in the two traditional areas encouraged the opening and extension of mines elsewhere. The production of high quality edge tools in Sheffield, famous from the Middle Ages, had long been based on local supplies of iron, and in the 1580s Lord Paget opened iron mines in Cannock Chase, Staffordshire, and built two furnaces and forges to manufacture the iron into goods needed locally. As with most mined materials, it was easier to process the minerals at the mouth of the mine rather than transport the low value, bulky materials to be manufactured elsewhere.

Non-ferrous metals

The total output of lead also increased at this time. Derbyshire remained one of the more important areas, and much was mined in the Yorkshire dales. The Mendip mines were extended, both over a larger area and deeper into the ground. The mines were divided into four districts, called mineries, each with its own town – East Harptree, Chewton Mendip, Priddy and Charterhouse. A lead reeve operated in each minery town, to collect the 10 per cent of smelted lead due to the landowner as a royalty, and to settle any disputes between the

16 Ocknells Manor in Berkshire, a house built in the early seventeenth century. The strength of the house comes from the timber frame that can be seen in places. Bricks have been used to fill the spaces, which is why there are a number of different patterns.

17 Several popular beaches in Cornwall and south Devon have dark caves in the cliffs. While some of them really were smugglers' caves, most are entrances or the drainage adits of tin or copper mines. This one is among three at the Lizard.

miners. Mining was also developed in mid-Wales, where a ton of lead contained up to 60 ounces (1,700 grams) of silver. Miners from Germany were brought over by Elizabeth I (1558-1603) to develop such mines in Cardiganshire, and she also had the mines at Combe Martin re-opened, though the best of the argentiferous lead had already gone.

Elizabeth was continually short of money, even before the war with Spain made matters worse. There were therefore obvious attractions in trying to increase the output of silver. Her encouragement of copper mining was based not so much on the value of the metal itself as on the taxes that could be levied on it and the possibility that miners searching for it might stumble on lodes of more precious metals. For this reason she gave Daniel Hochstetter, a man from Augsburg with considerable prospecting experience, sole right to search for copper and to mine it in Cumbria, in the hope that he could revive the industry there. It would be more correct to say she sold him the right, for he had to pay annually for the privilege, and any gold and silver found were to be Crown property. He started prospecting in 1564, and soon found copper at

Keswick and at nearby sites. He began mining under the title of the Company of Mines Royal in 1568, and had six mines open by 1570, which then produced 14.25 tonnes of copper, valued at the equivalent of £2,784.72. The ore was taken to Keswick for smelting. He had intended sending the ingots back to Germany, but when Elizabeth stopped that, he had the copper made up into pots and pans and sent those instead. Problems of drainage and the varying thickness of the lodes caused the mines to be running at a loss by 1600, and many years passed before they became profitable again. Some of the copper was taken by sea to Tintern on the River Wye, where Hochstetter had an interest in the Mineral and Battery Works. This specialized in making brass and iron wire, and the copper was alloyed with calamine (zinc carbonate) mined on the Mendips to make brass.

Copper mining was also revived in Cornwall during Elizabeth's reign, though by independent miners, not large companies. Mining for tin in Cornwall remained important throughout the century, and expanded a little towards the end. Polberro mine was one known to have been worked by 1600, and others are thought to have been, but accurate evidence is missing. Richard Carew described how the tin miners tested the value of a new lode in his *Survey of Cornwall,* written in 1602. The tools and methods mentioned were much the same for centuries before and after, and were broadly the same in all the mineral mines. He wrote:

> They sincke a shaft, or pit, of five or six foote in length, two or three foote in breadth, and seven or eight foote in depth, to proove whether they may so meete with the load . . . their ordinary tools are a pickaxe of iron, about sixteen inches long, sharpened at the one end to peck, and flat-headed at the other, to drive certain little iron wedges, wherewith they cleave the rocks. They have also a broad shovel, the outer part of iron, the middle of timber, into which the staff handle is slopewise fastened.

From this and other evidence, it can be reckoned that most mines were still shallow pits at this time because of the problems of drainage. The gradual increase in the demand for these metals encouraged miners to seek new sources of ore, but once these became flooded there was no easy way in which the work could be carried on. It may well be that some pits were drained by buckets passed up the shaft, or drawn up by a windlass. This was expensive, and only worth doing where the ore was rich – which was rare.

Coal

Drainage problems also affected coal mining, and prevented workings being more than day holes and bell pits. The problems might have been overcome had there been the need, but the demand for coal rose only slowly in the sixteenth century, mostly towards the end. Much of the coal produced came from the

18 Each clump of trees here has grown up on the remains of shallow coal pits near Flockton in Yorkshire. Such pits are called day holes or dene holes, and had to be shallow because of the risk of the sides caving in.

Tyne and Wear valleys, where the outcrops of coal in the valley sides made it easy to drive adits into the seams. Adits provided drainage, and also made it easier to take the coal out of the ground. Transport was also easy on the rivers, and most of the coal mined in the area was sent by ship to London, where it was known as sea coal. The shortage of wood and charcoal became serious in London about 1550, and it gradually forced soap boilers, dyers, brewers and salt refiners to use coal for heat instead.

Tyne and Wear was not the only place where coal was mined, though it was the largest because it was the only area that could send coal so cheaply to London. The other mines supplied customers over a much smaller radius. Mines in Broseley, Shropshire, produced a coal that contained little sulpher, and so could be used to heat iron for forging. It was sent via the Severn to Bewdley, where many blacksmiths worked, and was also used in the home by people within reach of the river trade. In West Yorkshire, coal outcropped on the steep valley sides, and coal was taken from bell pits at Flockton Edge in 1515, and from many other places from time to time. Some of the pits had short galleries extending from the bottom, exactly like Grimes Graves. Some of the coal was used by local ironworks, most of the rest by other industries. The Kay family, who bought lands at Honley near Huddersfield from the Crown after the closure of the monasteries, continued to mine coal from pits the monks had started, using it to burn lime to make fertilizer for the fields and in a smithy they built in 1573. By the end of the century, ships that took copper ore from Cornwall to South Wales for smelting returned with coal, most of which was again used for lime burning. Coal was also mined wherever it outcropped on the surface, and was put to local uses. The shortage of timber was mainly confined to the London area; but as the shortage began to be felt in other districts, the demand for coal increased more rapidly.

Chapter 4 A Time of Expansion

The era of mining expansion in the seventeenth and eighteenth centuries, was a time of considerable upheaval in Britain as a whole. At first this was caused by the bickering between kings and their parliaments, culminating in the Civil War (1642-9). The war was followed by a period of uncertainty (the Commonwealth), before Charles II was finally restored as head of the country in 1660. But friction still existed, and found expression in religious differences, leading to the overthrow of James II and the accession of William and Mary in 1689. Their arrival marked the beginning of a series of major foreign wars which did not end until 1815. Further tensions were caused by the rapid growth of the population, which increased by 50 per cent between 1760 and 1801. Such turmoil was bound to have an effect on manufacturers, and therefore on those supplying them with fuel and other raw materials. We will consider the period up to 1660 first, and then look at the period of more rapid expansion between 1660 and 1800.

Stones, 1600-60

The fashion for building or rebuilding manor houses, town houses and public buildings increased the demand for stone during the reign of James I (1603-25), and many new quarries were started. Most of these were small, some only producing enough stone for one building. The best stone was worth transporting some distance, causing larger quarries to be developed, especially in the Cotswolds and Northamptonshire. Slates were easier to move than blocks of stone and the demand for them continued to grow as more householders realized the dangers of thatch as a fire hazard. Several quarries in Cornwall supplied slates to many parts of Britain – those at Delabole sent large quantities from the harbour at Port Gaverne to Brittany and the Netherlands as well. Slates were also dug in North Wales and the Lake District, though not to the same extent as in Devon and Cornwall. The need for clay to make bricks and tiles grew even faster than the output of the stone quarries, because of their popularity and convenience in areas without good building stone. Sand was also needed to make bricks (and a revival of glass-making led to an increased use of sand as the century passed). Bricks and stones were all laid in mortar, which was made from the lime obtained from limestone. The limestone was burned in open kilns to make it suitable for mortar, and lime was also increasingly used as a fertilizer. This all added up to more work in the limestone quarries.

Base metals, 1600-60

Development of non-ferrous mining was uneven in these years. The copper mines, both Hochstetter's and others, were scarcely worked at all, because it was cheaper to import refined copper from Sweden. The Swedish miners had more accessible lodes of ore and endless supplies of charcoal for smelting it; the shortage of timber for charcoal in Britain made smelting expensive and attempts to use coal instead were unsuccessful. The Mines Royal combined to produce a little copper and some was mined in Cornwall and sent to South Wales for smelting, but all the mines had to struggle to make sales match costs.

The Cornish tin mines were free of foreign competition, and output continued to increase in the first half of the seventeenth century. The demand for tin to make pewter grew quite rapidly, as people changed from wooden plates to the more graceful shapes that could be beaten from pewter. Skilled tin miners were paid up to 8d. (3p) a day in 1602, at a time when farm workers were lucky to earn a penny. The demand for lead, the other ingredient of pewter, also grew, and mines flourished in most of the traditional areas. The Mendip mines, hardly any larger than in pre-Roman times, had their last years of profitable working in the period up to 1680, and supplied lead to all parts of south-west England, London and western Europe. Much of it went through Bristol, the second largest port in Britain. In 1608, for example, 1,069 tonnes of lead left Bristol for Hamburg, Jersey, Bordeaux, Cadiz and Madeira. (By comparison, 30 tonnes of iron were exported that year, 40 tonnes of iron ore and 20 tonnes of coal.) Most of the trade was in small quantities (as were most goods until the late eighteenth century) as the following entry from the port book for 1623 indicates:

> In the *Prymrose* of Brightemstowe, Nicholas Payne master, for Dartmouth Richard Hallworthy – xl barrels of lead Oare and iiiC lv pieces of lead in sowes.

Mining techniques in Cardiganshire were somewhat more advanced, encouraged by the setting up of a mint at Aberystwyth in the 1640s. German, Derbyshire and Yorkshire miners were brought in to dig shafts and extend the galleries, using their knowledge of wooden pit-props. The rock was loosened in the galleries by fire-setting: a roaring fire was lit against the face to be broken to heat the rock, and cold water was dashed on to the rock when it was hot. The sudden contraction made it fracture and, when the smoke, fumes and steam had thinned, miners went in with hammers and wedges to break the stones from the mass of loosened rock. Fire-setting added to the problems of ventilation and drainage. Adits were dug to drain some mines, while Hugh Middleton, a mine engineer who sank several shafts in Cardiganshire before 1620, used gangs of men to pump the water vertically from trough to trough up the shaft. Hand bellows were tried to blow air through some mines, but it was more usual to have a man flapping a coat to keep the air moving. Far more ambitious drainage schemes were started in Derbyshire. Cornelius Vermuyden, a Dutchman who knew much about fenland drainage, supervised the digging of a long adit called Cromford Sough. This ran under many mines in the Peak District, and so underdrained them, making deeper mining possible. The output of lead also increased in the north Pennines, and new mines were started in Galloway.

Iron, 1600-60

Iron mining also increased in the first half of the seventeenth century, especially during the war years (1642-9). Much of the mining was still done by individuals or small partnerships, though there were a number of men prepared to invest substantial sums of money in this industry. William, Earl of Pembroke, for instance, was given a grant in 1612 of:

> 12,000 cords of wood yearly for twenty-one years at 4s. [20p] per cord, being £2400 . . . with liberty to dig for and take within any part of the said Forest . . . so much mine ore, cinders, earth, sand, stone, breaks, moss, sea coal and marle as should be necessary for carrying on the ironworks let to him, or which he should erect.

Later in the same year he was given 'the lordship, manor, town and castle of St Briavels, and all the Forest of Dean with the appurtenances, and all lands, mines and quarries belonging thereto . . . at the yearly rent of £83.18s.4d. [£83.91½]'. This was a large sum of money, even for a whole forest, and the Earl intended to pay it out of the profits of the iron mines and furnaces. Larger and more efficient blast furnaces were later built in the Forest of Dean, and also in Sussex and most other mining areas, which greatly increased the demand for iron ore. Possession of the mines and furnaces was important during the Civil War. The Royalists kept control of the Forest of Dean until their defeat in 1646, and in 1650 Parliament ordered the furnaces to be shut down in case the Royalists should recapture the area. They did not, but the furnaces were not restarted until 1660.

Similar violent swings in the demand for iron ore occurred in all the mining areas during and after the war. In parts of the Midlands, upsets were caused by the shortage of fuel for smelting. All the timber around Wednesbury (Staffs), where iron had been mined since the thirteenth century, had been used by the middle of the seventeenth century, so the ore was taken by pack animal to the RiverTame and by boat to Aston and Perry Barr, where there was both wood and water power for the furnace bellows. Local iron manufacturers had their own mines, and employed as many miners as fitted in with their plans. John Jennens, who lived at Wednesbury Hall, had iron mines nearby and ironworks in Aston.

Coal, 1600-1800

Expansion in all the metal mines, however, was nothing compared to the growth of coal mining in the seventeenth and eighteenth centuries. From being a neglected industry, suitable in the Middle Ages for little more than ballasting empty ships, it began to emerge as one of the chief assets in the development of British industry. There are few statistics from the early period, and none that accurately record production for the whole country. The nearest we can get are the figures for the coal trade from Newcastle to London. This jumped from 32,424 tonnes in 1563-4 to 520, 567 tonnes in 1658-9. Most of the coal mined in the Tyne and Wear valleys went to London, but the amount going to other European countries also increased rapidly. Though figures are scarce for other coal producing areas, there are signs that output increased there too, and that new markets were being found for it. The 20 tonnes exported from Bristol in 1608-9, which had no doubt come from mines around Kingswood and in north Somerset, went to customers in Amsterdam, Nantes, La Rochelle and Bordeaux.

Most of the coal raised, however, was used in Britain, and the shrinking supplies of charcoal forced manufacturers to find ways of substituting coal. Some processes were simple to change: the boiling of salt or burning of lime and bricks required heat, but it made no difference what caused that heat. By the beginning of the seventeenth century, coal had already replaced charcoal (in London and the coal-mining areas at any rate) for brewing, distilling, boiling sugar and soap, firing pottery and heating iron for rolling into sheets. Coal could not at first be used for some other processes because the impurities in it contaminated the product. Charcoal was used to melt glass until about 1610, when the practice of sealing the ingredients in a closed crucible made it possible to use coal. Fumes prevented coal from being used to dry malt until it was found, about 1645, that if the coal was made into coke first there were fewer fumes and no smoke.

By the end of the seventeenth century, coal was being used to smelt lead in a reverberatory furnace, which avoided direct contact between the fumes and metal, and the same method was used for tin and copper early in the eighteenth century. Industrial development had been a little stunted by the shortage of charcoal in the sixteenth century; now the use of coal boosted industrial output and so created an increasing demand for it. There was plenty of coal available, and production was estimated to be about two million tonnes by 1660, five times the output of the rest of the world combined at that time.

Much of the coal came from very small pits, and nearly all the mines were open pits or bell pits, except in the more intensively mined north-east. Manufacturers frequently sank their own mines as a private and reliable source of fuel. In Scotland, salt pans and coal mines were often run together. A Lancashire bricklayer mined coal at Farnworth in 1647, while in Shropshire, a nailer called John Bennet rented an acre of pasture which had coal works, iron pits, fireclay and limestone – all for 3s.4d. [17p] a year. Some shafts and

19 A coal mine at Broseley, near Coalbrookdale in Shropshire. A two horse-power gin is used to lift the coal up the shaft, and the tall chimney suggests the use of a steam pumping engine, though there is no sign of one. Notice the pack animals, wagonway and drying house for the miners.

mines were dug in the Midlands after 1650 to cope with the rising demand, but small day pits were more common.

The demand for coal grew steadily between 1660 and 1760. Expansion in other industries increased the need for fuel, particularly the use of coke for smelting iron and in other processes. The growth in the population also meant that more was needed for cooking and heating. Though the methods used did not change much (see Chapter 5), the number of pits and the numbers employed in each steadily increased. There were 70 mines in Kingswood Chase in 1684, but they only employed 123 men between them. The 90 pits in the Forest of Dean employed a total of 662 in 1778, but the weekly output scarcely averaged 20 tonnes per pit. Mines in the north of England were larger. A Whitehaven pit employed 19 in 1675, and 40 was about average by 1750. Many shafts in Northumberland and Durham were over 60 metres deep at the beginning of the eighteenth century, and some as much as 120, but few galleries extended more than 150 metres from the bottom of the shaft because of limits put on by the owners, the problems of ventilation and drainage and the fact that the barrow-men were paid piece rates. These factors combined to make it necessary to sink new shafts quite frequently.

The use of more and larger pits made it possible to expand production from about 2.5 million tonnes in 1700 to about 6 million in 1770. Some of the larger mines were worked by partnerships of men, who between them could find the capital needed to finance the digging of deeper shafts, or installing the expensive atmospheric engines designed by Thomas Newcomen to pump out mines used at a few Midlands pits from 1712. The Grand Allies was the most wealthy of the partnerships: George Bowes, Charles Montague and Colonel Liddell went into partnership in 1726, and controlled many of the most productive mines in Northumberland and Durham. Such groups were exceptional, though, and were mainly found in that area. Elsewhere, most mines were small, individually owned, and worked by a handful of men.

Unrest among miners

Miners around the country had a great deal in common, and were ready to join together if circumstances encouraged them. There was widespread rioting among miners whenever the price of wheat rose, as it did in 1709, 1727-8, 1740 and 1756-7. Eight cornmills were destroyed around Nottingham in 1756, (which cannot have reduced the prices), and an anonymous writer who witnessed the rioting in Leicestershire commented on it and on the increasing importance of coal to industry.

> I need not observe that ye Circumstances of Colliers are very different to any other men not only as they all act in League and would stand by one another thro'ought the Kingdom, and are desperate fellows (which is seen by their attacking Gaols to release any that are confined) but besides this they think they can at any time hide themselves and they know that ye Kingdom cannot do without Coals and they know that other People can't do their Work.

The metal miners also rioted occasionally (2,000 miners armed with clubs ransacked granaries in Falmouth in 1727) but their industries were not expanding nearly as fast as coal mining, and their wages and working conditions were normally somewhat better. Iron mining in Sussex declined as reserves became exhausted, while iron ore mined in the Forest of Dean had to be taken to furnaces outside the forest after 1674 because those in the forest were pulled down in a bid to save wood. Some of it went via the River Severn to the Midlands, where iron mining and working quickly expanded. This increase was particularly rapid after about 1760 when Abraham Darby's use of coke to smelt iron became common knowledge. Iron mines were also set up in

Lincolnshire and Northamptonshire, and in Scotland and southern Ireland. Most of the mines though, continued to be small and scattered.

Base metals, 1660-1800

The copper mines in Cornwall were rather more ambitious than the coal mines. The two sixteenth-century companies combined in 1668, and the Mines Royal Act in 1689 freed the owners of tin, lead, iron and copper mines from the heavy hand of royal right to any gold and silver found in mines. John Coster set up a new copper smelting works at Redbrook on the Wye, and the English Copper Company set up close by in 1691 and was processing 1,000 tonnes of Cornish ore a year by 1700. Smelting works at Neath were enlarged in 1695, and new works set up in Swansea in 1717, the year in which copper coinage became part of the legal currency. The increased demand for copper, as sheet and to make brass, was reflected in a considerable growth in the output of ore in Cornwall.

John Coster was responsible for much of the increase, for he introduced many new mining methods into the county. Among these were the use of water wheels for drainage, the cutting of drainage adits to save the cost of lifting water to the surface, and the construction of the first horse gin which increased the depth from which ore could be raised. Gunpowder was used in some mines before 1700, but fire-setting continued in most well into the eighteenth century. Coster was not the only innovator. The first Newcomen engine was set up in Cornwall in 1720, and many more were built after 1741 when the duty was abolished on coal carried around the coast. Most of the copper was raised from the Redruth-Camborne area, and the output increased from 6,000 tonnes in 1720 to 11,000 in 1739 and 29,000 in 1770. The value of copper mined exceeded that of tin in 1740 and continued to rise. There were 20 major mines in production by 1770, and 70 smaller ones. The demand for tin did not increase at anything like the same pace, though the Cornish mines did expand a little and the richest of them, Wheal Vor at Breage, could afford to install an expensive Savery steam pump.

The Mendip lead mines, which had once been so rich, began a long, slow decline at the end of the seventeenth century. Most of the lead had outcropped near the surface and, although some ore was obtained by deeper mines, it was not much and the mines were primitive. Elsewhere, the Mines Royal Act had a tonic effect. Mining began again in Flintshire after a long gap, and workings in Cardiganshire were extended. The area of greatest growth was Derbyshire. Gunpowder was used in a few mines there from 1692, to loosen the rock. Mining in the Yorkshire dales and North Pennines also increased but was handicapped by the lack of suitable transport – pack animals were the only way of moving both ore and pigs of lead in most parts until the nineteenth century. Further north again, mining was extended in the neighbourhood of Leadhills and Wanlockhead.

Stones, 1660-1800

The need for building stone increased with the size of the population and as it became fashionable to rebuild country mansions in the classical style. The largest single demand came from the rebuilding of London after the destructive fire of 1666. As it was now possible to cut Portland stone with water-powered frame saws, Christopher Wren made free use of it in his designs – more than a million tonnes were sent to build St Paul's Cathedral and the other churches of London.

Clay pits were extended to provide material for bricks and tiles, and the extraction of ball clay (sometimes called pipe clay) also increased rapidly. This was used to make tobacco pipes in the seventeenth century and later to make

20 *Opposite* Men shaping blocks of Portland stone in the quarries in Dorset. Notice the size of the blocks, and the range of hand tools with which they are being cut to shape.

21 *Below* Cutting ball clay in the 1920s, by methods which had hardly changed in 200 years. The clay was stiff enough to be cut in blocks, and the wagonway took the balls to the canal and so to Teignmouth.

pots and plates in the Staffordshire potteries. The clay was only found in Devon and Dorset. A traveller, Jeremiah Milles, visited some of the Devon pits about 1750:

> Kingsteignton. On ye north side of the parish and of ye high road from Exeter to Newton in some coarse fields called Bellamarsh are pitts where they dig great quantities of ye finest pipe clay. There are three of them now in work not far from one another. The surface of the ground where this clay is found is a black moorish heathy soil for about a foot or two perpendicular, below this, via yellow coarse clay which grows brighter and brighter till it comes to ye depth of about 12 feet [3.6m.] and then succeeds a bed of perfect white and clean clay, without ye least mixture of stone or sand. It is of a very soft greasy nature, and it is not only used in ye country to make pipes . . . but there are great quantities of it continually carried to Staffordshire and it is found very useful in making their earthenware. The land where the clay is dug is part of the Manor of Preston. The Tenant

22 A nineteenth-century impression of an alum mine. The methods are the same as most kinds of mining, though the lack of dangerous gases made it safe to use ordinary candles. Few methods changed in the mines from the sixteenth century to the nineteenth.

generally employs Labourers to dig it at the rate of 1s.4d. [7p] per Tun. They cutt it into square pieces about one foot long and nine inches broad and as many thick and of about 35 pounds [16 kilos] weight each. In this manner they load it on horseback and carry it two miles to a place called Hackney in Kingsteignton parish where it is shipped on board vessels for Liverpool from which place it is carried to Staffordshire. The owner of the land sells it at ye waterside for about 7s. [35p] per tun.

The methods described here for extracting clay from open pits and transporting it to customers were similar to those used by the miners of most other surface deposits.

Alum was mined on the Yorkshire coast near Scarborough, using coal brought from Newcastle to boil the alum solution. This was one of the many chemicals needed as industrial processes became more complicated. Another example was the oil shale quarry started in 1697 by Martin Eele at Broseley in Shropshire. He also had a coal mine nearby, and he used the coal to extract oil, turpentine, tar and pitch from the shale. The small tonnages of such commodities produced were often vital to other industries.

Chapter 5 When Coal was King

Coal was being raised from Pentre Colliery near Swansea from about 1745. Gruffydd Price was working the pits in 1768, when he installed a steam pumping engine. He continued to mine coal until 1787, raising about 10,000 tonnes a year, but lost money for the last three years. He decided to cut his losses by leasing the colliery to a firm of copper smelters in 1788, who had the capital to sink deeper shafts and install a more powerful engine. Output soon exceeded 16,000 tonnes. This use of profits obtained from metal working to extend coal mines is an example of the close links between the mining and manufacturing industries. In this chapter we are going to look at the coal industry and the problems faced in the coal mines, and in Chapter 6 we will compare these with mining for metals.

Rising demand

The demand for coal began to increase rapidly towards the end of the eighteenth century, and the rate of growth continued to accelerate until the First World War (1914-18). This was mainly due to the expansion of industries at this time – the new cotton mills, for example, needed vastly more fuel than in the past, to heat the water for washing, bleaching and dyeing the cloth, and to dry it. Above all, the industry needed power, which was readily available after 1781 when James Watt made his improved steam engine capable of driving other machines. The power and use of the steam engine increased dramatically after 1800, and steam power was used in many industries, not least in coal mines. One colliery owner, R. Protheroe, had 32 pits in the Forest of Dean and reported in 1832 that:

> The depth of my principal pits at Park End and Bilson varies from about 150 to 200 yards [130-180m.]; that of my new gales, for which I have engine licences, is estimated at from 250 to 300 yards [230-270m.]. I have 12 steam engines varying from 12 to 140 horsepower, 9 or 10 of which are at work, the whole amounting to 500 horsepower; and I have licences for four more engines, two of which must be of very great power.

The development of railways soon followed, providing not only a great demand for coal for themselves but also supplying industrialists in many areas with the opportunity to make use of the fuel. Engines of over 2,000 horsepower were being installed in textile mills by 1900, and multiple expansion engines were being used in ships. Quite apart from industry, the continuing increase in the population and their rising standard of living brought a substantial demand for coal to heat homes and cook meals. London and some of the towns on coalfields had gasworks by the 1820s, to provide street and house lighting. Gas (derived from coal) became more popular after 1860, and was used for heating, cooking and as a source of power for industry; electricity too was obtained from coal from the 1880s onwards.

The expansion of coal output was dramatic. Production was estimated to be 2.5 million tonnes in 1700, 6 million in 1770 and 10 million by 1800. Accurate figures were collected from 1854, which show the rapid expansion of the industry in the nineteenth century, especially in the Midlands and South Wales.

Average annual output per decade, in million tonnes								
	North-umberland & Durham	Yorks	Derbys Leics Notts Warks	Lancs & Cheshire	Staffs & Worcs	South Wales	Scot-land	Total
1854-63	17	9	5	11	7	7	9	72
1864-73	26	11	9	15	13	13	14	107
1874-83	33	17	14	21	14	19	19	139
1884-93	36	20	18	24	14	27	23	167
1894-1903	44	25	26	27	14	36	30	207
1904-13	52	35	34	27	14	48	39	256

The figures given as totals include all the smaller mining districts, such as Somerset, Cumberland, Cork, and, from 1912, the new field in Kent. The amount of coal exported also grew, from about 12 million tonnes a year in 1870 to 95 million in 1913, the highest figure ever reached. Most of the exported coal went to countries in Europe, and the rest was sent to maintain coaling stations for steam ships on the ocean routes.

Colliery owners

All the coal mines were privately owned in the nineteenth century, but there was a wide range of owners. Some were the landowners of thousands of acres, such as the Duke of Hamilton in Scotland, the Duke of Norfolk in Yorkshire and the Duke of Devonshire in Derbyshire. It had always been accepted that anything in the ground belonged to the owner of the surface soil. In the

23 *Overleaf* A somewhat horrific engraving of the Black Country, near Wolverhampton, made in the mid-nineteenth century. One-horse gins are still the common kind of power used for bringing coal to the surface, and steam engines for pumping. The mass of smoking chimneys in the distant towns indicate the growth in the number of customers.

24 A row of colliers' cottages at Long Benion. The toilets draining into the stream, the single water pump, the unmade road and general air of dereliction make a dismal scene.

eighteenth century, the titled landowners were able to use their power in Parliament to prevent any increases in duties on coal, so helping to boost output still further. Some landowners, like the Duke of Bridgewater, took an active interest in the collieries on their land, but most preferred to employ a manager or lease the mines to someone else to work. At the other end of the scale, many small landowners also mined coal from their own fields. This might only be sufficient for their lime-burning needs, though some raised enough to sell to their neighbours as well. Between these two extremes were the owners of country estates, some of whose income came from the coal under their land. All landowners took a royalty on the coal removed from their land, which usually amounted to 10 or 12 per cent of its value. Some also imposed conditions on those working the mine, such as a limit on the quantity to be mined a year or the removal of all spoil tips when a mine was worked out.

The Birmingham Coal Company started in 1793 as a co-operative venture with more than 100 members. They jointly raised the thousands of pounds needed to develop new mines, and shared the profits they made. The company ran for many years, though the co-operative idea was not copied. It became increasingly common after 1850, however, for coal to be mined by companies rather than by individuals. The cost of searching for coal, and of draining and ventilating pits that grew steadily deeper, put coal mining beyond the means of most individuals. Some went into partnership with a few others, and the practice of selling shares to the public began to gather pace from the 1890s. One permanent feature of coal mining was that costs of production were always increasing – every tubful of coal removed left the next that much further away. Such costs as transport, lighting, ventilation, drainage and haulage were bound to increase rapidly with the hectic pace of development in the nineteenth century. Indeed the demand for coal was so great that many narrow seams were worked, and this was expensive. As a result, the annual output per man actually fell from 397 tonnes in 1881 to 326 in 1901 and 304 in 1911. Though this increased the price at which the coal was sold it did not seem to matter, for the coal companies had no difficulty in selling all they could mine, as the output figures indicate. It was bound to cause difficulties, however, should any cheaper fuels become available.

The size of mines varied greatly, depending partly on the owner and manager who had to decide how much they could expect to sell, and partly, of course, on the thickness of the seams. There could be a wide range of sizes even in a small area, where the collieries were working the same seams. There were 76 collieries in Glamorgan in 1842, to take one example: 16 produced less than 100 tonnes per year, 35 raised 2,500-15,000 tonnes, 20 brought out 15,000-35,000 tonnes and one mined more than 70,000 tonnes. Such variations were repeated in all the mining areas and, although the annual production figures steadily rose, this remained the pattern in the industry until 1914.

Mining communities

Many coal mines were remote from towns and villages, so mining villages grew up around the larger collieries. All too often, because they were erected quickly, these consisted of drab rows of houses, as if permanently blackened by the coal dust. But they were real communities, where the families frequently had to help each other – through grief after accidents, and poverty when prices fell. Equally, they knew how to celebrate. Mining villages were frequently avoided by outsiders in the eighteenth century, because of the miners' reputation for violence. This was often true enough. Kingswood miners took protection money from shopkeepers in nearby Bitton, and carried out periodic raids on shops as far away as Hereford. But there was a positive side to the mining villages too. George Stephenson lived in a mining village until he started travelling to build railways, Victor Purdy in Kingswood wrote 2,000 hymns and, at Mangotsfield, William Fluelling ground his own lenses to make telescopes and became a knowledgeable astronomer. Mining communities were not inhabited by rogues and drunkards, but by groups of families who had to accept the possibility of sudden injury and death as a hazard of everyday life.

Mining methods

The first stage in opening a mine was to bore into the ground to find out how deep the coal was and how thick the seam. This had to be done with hand tools in the eighteenth century, and required considerable effort. As a result, the borers were highly paid. The usual rates were 5s. [25p] for the first five fathoms [about 10 metres], 10s. [50p] for the next five, 15s. [75p] for the next and so on. If the bore showed that it was worth mining, borers were replaced by sinkers. Their job was to cut a shaft down to the seam. From 1719 they had the help of gunpowder for this job, but much of the work still had to be by pick, shovel and crowbar. Square shafts were cut through the layers of soil, and timber was used to shore up the walls. The shaft became circular as soon as the sinkers reached solid rock. They were also paid on a rising scale: at Griff colliery,

25 Two Staffordshire miners preparing to go down a mine. Note their dress, and the flask with a drink.

Warwickshire, in 1701 this was 2s.6d. [12½p] an ell [1.23 metres] for the first 4 ells, 3s. [15p] for the next four and so forth. Boring and sinking went on continuously in the eighteenth century because, with shallow shafts, it was cheaper to sink new ones than to haul coal through long underground galleries.

There were two main methods of mining the coal underground in the eighteenth century, called 'long-wall' and 'pillar-and-bord'. Both methods, with some local variations, were still in common use in 1914. Pillar-and-bord was the more traditional method, used in the eighteenth century in Northumberland and Durham, Scotland, Cumbria and most of the other areas where large quantities of coal were raised. The sequence of operations in opening up a new mine was begun by digging a tunnel from the base of the shaft left by the sinkers. The tunnel went through the coal in whichever direction it cut most easily, and was made 1½ to 2 metres high. This became the main heading – the road along which all the coal would travel to the foot of the shaft. Other tunnels (called galleries) were cut at right angles to the main heading at intervals of about 50 metres. Up to this point, the miners all worked together to create the system of underground roads, and were paid a common wage. Now each man took his place along a gallery and began to cut his bord – the section of seam which only he worked. Bords were three metres wide, and each was separated from the next by a pillar or wall of coal four metres thick, which was left standing to support the roof. The miner cut his bord for up to 40 metres then another thick pillar was left, a new gallery cut and a fresh set of bords started. (The widths varied from one mine to another. Thicker pillars had to be

26 A solitary miner working in his bord. A pick and shovel are his only tools, and the light from a safety lamp left many shadows. The glow in the distance from the next bord down the gallery indicates the thickness of the pillar.

left if the coal was crumbly or the gallery was deep, but the pillars could be reduced if the coal was sound. It was always a temptation to nibble away at them, especially when bords were almost finished.)

Each miner was in charge of his own bord. He was the hewer, who cut the coal with a pick and employed his own workers to take the coal up to the surface, or to the bottom of the shaft in some mines. The hewer was either paid by the quantity of coal he produced, or received a standard wage provided he mined an agreed quantity of coal a week. Those he employed were usually members of his own family, to whom he paid very little. The custom of a whole family unit working together was similar to domestic industry on the surface, but the physical effort was tremendous. Family involvement was particularly common in Scotland. The usual practice there at the beginning of the nineteenth century was for the men and their sons to go to work at about 11 o'clock at night:

> In about three hours after, his wife (attended by her daughters, if she has any sufficiently grown) sets out for the pit, having previously wrapped her infant child in a blanket, and left it to the care of an old woman, who for a gratuity, keeps three or four children at a time, and who, in their mother's absence, feeds them with ale or whisky, mixed with water . . .
>
> The mother . . . descends the pit with her older daughters, when each, having a basket of a suitable form, lays it down, and into it the large coals are rolled; and such is the weight carried, that it frequently takes two men to lift the burden upon their backs: the girls are loaded according to their strength. The mother sets out first, carrying a lighted candle in her teeth; the girls follow, and in this manner they proceed to the pit bottom, and with weary steps and slow, ascend the stairs, halting occasionally to draw breath, till they arrive at the hill or pit top, where the coals are laid down for sale; and in this manner they go for eight or ten hours almost without resting.

The basket was called a creel; it was held on by a strap round the forehead and contained up to 77 kilos. Some of the journeys were 150 metres along the gallery and heading, 35 metres up the ladders in the shaft and perhaps 20 metres to the surface dump, and the carriers made 24 trips a day. Wives and daughters in other mining areas were not expected to do as much work as this, but hewers in all parts relied on their children to take the coal to the shaft, either in baskets, on sledges or in the wheeled tubs that came into use in the eighteenth century. Pillar-and-bord working was normal in most areas, and was adapted to local conditions. It was called stoop-and-room in Scotland, and stall-working in South Wales where the thin seams made it necessary to take out wider bords. The disadvantage of the method was that anything up to 60 per cent of the coal had to be left as pillars to support the roof. This meant that more time was spent cutting galleries and carrying the coal along them than actually winning coal.

Long-wall working was developed to overcome this shortcoming. The method originated in Shropshire in the seventeenth century and spread to many other areas in the eighteenth: by the end of the nineteenth century, Durham was the only major area still practising pillar-and-bord mining. Instead of each hewer working on his own, long-wall working involved teams of men, each specializing in its own kind of work. The method is explained in this report of 1811 from Derbyshire:

> The working . . . commences by a set of Colliers called Holers, who begin in the night, and hole or undermine all the bank or face of the Coal, by a channel or neck from 20 to 30 inches [50-70 cm] back, and 4 to 6 inches [10-15 cm] high in front, pecking out the holeing-stuff with a light and sharp tool called a pick, hack or maundrel: and placing short strutts of wood in such places where the coal seems likely to fall, in consequence of being so undermined . . .

27 Long-wall working. The two hewers are undercutting the seam, temporarily held up on the short supports at each edge of the photograph. These will be removed when the cutting is complete, and the weight of the coal will bring it down.

When the Holers have finished their operations through the whole length of the Bank, or Banks, and cut a vertical nick at one or each end of the Bank, called the cutting-end, and have retired, a new set of Men called Hammer-men, or Drivers, enter the works, and fall the Coal, by means of long and sharp iron wedges, set into the face of the Coal at top or near it . . . which they drive by large Hammers, till the Coal is forced down, and falls in large blocks, often many yards in length . . . a man called Rembler next follows, and with a hammer-pick breaks the blocks of Coal into sizeable pieces: and the drawing apparatus being ready, the loaders fill the Coals into the Corves or Trams . . .

A new set of Men now enter the Pit, called Punchers or Timberers, taking with them a number of stout posts of wood, cut or sawed off to a certain

28 Though steam power was used in all areas for drainage and, later, lifting men and coal to the surface, it was impossible to use it below ground to cut coal. The 'Iron Man' (*top*) was worked by hydraulic power; its cutter arm cut the slot for long-wall working. Compressed air proved more successful. The disc cutter (*bottom*) was built in 1868 to do the same job, and was the first of a long line of machines powered the same way.

length . . . These puncheons they set up in a row, in front of or almost touching the new face of the Coal, applying a small flat piece of wood, or templet, at top of each, unless the roof which they punch-to, as it is called, be very hard . . . The work is now ready for the Holers to return, and after another day's work as above described, the Punchers return, and in pretty good roofs they take down the puncheons in succession and remove them forwards almost to the face of the Coal, as before.

The workings were pushed forward in a long line without leaving any pillars of coal. Stone and small coal (which had no market then) was packed into the empty space, leaving a roadway for the trams, and the roof allowed to fall on to it. This meant that the method could not be used under towns. Most of the coal mined in 1914 was still dug out in this way.

For a long time picks, wedges and shovels remained the most common tools; few mine managers were enthusiastic about adopting machinery. The first mechanical cutter, called Willie Brown's Iron Man, appeared in 1761. Though it was little more than a pick head made to swing by manpower, it could hit harder and faster than a hewer. Other kinds were designed early in the nineteenth century, worked by men or horses, but they were of little use. Harder metal for the cutters and a stronger source of power were needed. Steam, which was so changing the face of industry on the surface, was no help to miners below ground because of the danger of fire. The increasing power of steam engines, however, was a great help in increasing the depth from which water could be pumped, for hauling people and coal to the surface (from 1830, and from greater depths with wire cables in the 1840s) and to ventilate mines by suction fans (1850s). The first successful coal-cutting machine was devised by Thomas Harrison in 1863, by which time harder steels and compressed air were available. The narrowness of many coal seams, the small size of most collieries (meaning that they had few reserves of cash for investing in expensive equipment) and the lack of initiative of many mine managers combined to limit the use of machinery. Only 8 per cent of coal was mechanically cut in 1913, and about the same amount was handled mechanically on conveyor belts. Steel props, first used in Wales in 1894, replaced timber props just as slowly.

Children in mines

Changes in working conditions were brought about in a quite random manner. The young age at which some children went down the pit, for example, had attracted comment even in the eighteenth century, though it was thought that parents knew best what was fit for their children. Few critics realized that the low price of coal made for low wages, and that miners had to supplement their own wages with the pence their children earned. Many children started work in mining areas at the age of 10, some at 8, and they worked a longer day than the hewers. In the larger mines, with rambling galleries to ventilate, there was work for even younger children as trappers, shutting the gates after trams had passed to keep the air current going around the workings. At length, in 1840, a government commission was appointed to enquire into the working conditions of those under 18. Their report in 1842 showed that in most areas children began working at 7 or 8, and some even younger. William Richards, interviewed at Buttery Hatch Colliery in South Wales at the age of 7, had started in the mine at 4: 'When I first went down I couldn't keep my eyes open; I don't fall asleep

now, I smokes my pipe.' In Scotland and a few other areas, however, children did not start work until they were 12. The youngest children were trappers, of which the report said:

> Although this employment scarcely deserves the name of labour yet, as the children engaged in it are commonly excluded from light and are always without companions, it would, were it not for the passing and repassing of coal carriages, amount to solitary confinement of the worst order.

They worked from 12 to 18 hours a day for sixpence [2½p]. At 12, they became hurriers, dragging the tubs of coal to the bottom of the shaft: they toiled for the same hours but for ninepence [4p] or a shilling [5p] a day. At 14, they were put in charge of a tram, and were called putters. The next promotion, at 15, was to put-and-hewer, at 2s. [10p] to 2s.6d. [12½p] a day, and finally to hewer at 17 or 18. The hewer worked only 8 to 10 hours, often for more pay. This usually depended on how scarce miners were in the area. In

29 *Opposite* The search for coal in the nineteenth century called for increasingly more powerful drilling machinery. This drill, whose cutting head was fitted with diamonds to break through rock, was used to bore samples in Kent in the 1870s. Coal was eventually found below 1,000 metres but was not mined until the twentieth century.

30 *Below* One of the dramatic illustrations in the 1842 report on children in mines, which so horrified parliament.

1. The candle-holder: a socket of iron, having a spike at right angles for the convenience of sticking the light in the sides of the pit when stationary. The spike also forms a handle when the light is carried before them.
2. A skull-cap, having a leather band, into which the candle-holder is thrust when the hands are employed in locomotion.
3. The girdle and hook for attaching to the chain.
4. Represents the position of the girdle.

Llanelly, for instance, colliers were paid 20s. [£1.00] to 30s. [£1.50] a week in 1807, but only 1s.6d. [7½p] to 1s.8d. [8p] a day in Pembrokeshire seven years later. There was no national, or even local, agreement. Some miners were paid in truck (that is, they were paid in goods instead of money, or in money on the understanding that they would buy anything they needed from their employers), which was not finally stamped out until the 1840s. Others received a minimum wage each week and the balance every three or four months. However, colliers were regarded as better off than industrial workers, and Lancashire mill lads muttered:

> Collier lads get gowd and silver,
> Factory lads get nowt but brass.

There were other advantages too, of which the greatest was the supply of free or concessionary coal, which guaranteed a warm home, warm meal and dry clothes the next day – comforts which were often denied to other workers.

Working conditions

The working conditions of coal miners were dangerous, with death and injury an ever-present likelihood. Several gases collected in most coal mines. In the shallow workings common before 1760, choke damp (carbonic acid gas) accumulated. This suffocated miners because there was insufficient oxygen in the air, but it usually put the candle out first, giving a warning of danger. A greater menace was fire damp (methane gas and air), which a candle exploded. Fire damp frequently collected in the old workings of deeper mines, or even overnight in sections being worked. The *Annual Register* recorded in 1773:

> Dec. 6th. The foul air in an old waste of a colliery near the River Wear in Yorkshire took fire, and breaking down the barrier or partition between the waste and the working pit, made the most terrible explosions ever beheld. The pit is said to be 80 fathoms deep; and everything in the way of the blast was thrown out at the mouth to the estimated height of 200 yards in the air. Most of the pitmen, having just in time discovered the danger, were drawn up, and escaped unhurt; but some boys and one man who were left behind lost their lives.

Explosions were so frequent in the eighteenth century that they were usually ignored – no inquests were held on pitmen killed in the north of England before 1814, and there was not even a coroner in Scotland until 1842. Miners were usually just as indifferent: a serious fire at Tanfield Colliery in 1740 developed from a fire the miners lit in the colliery to warm themselves. The answer to all the gases was adequate ventilation, but this was seldom easy to arrange. Natural air currents would flow through a series of adits and shafts, but

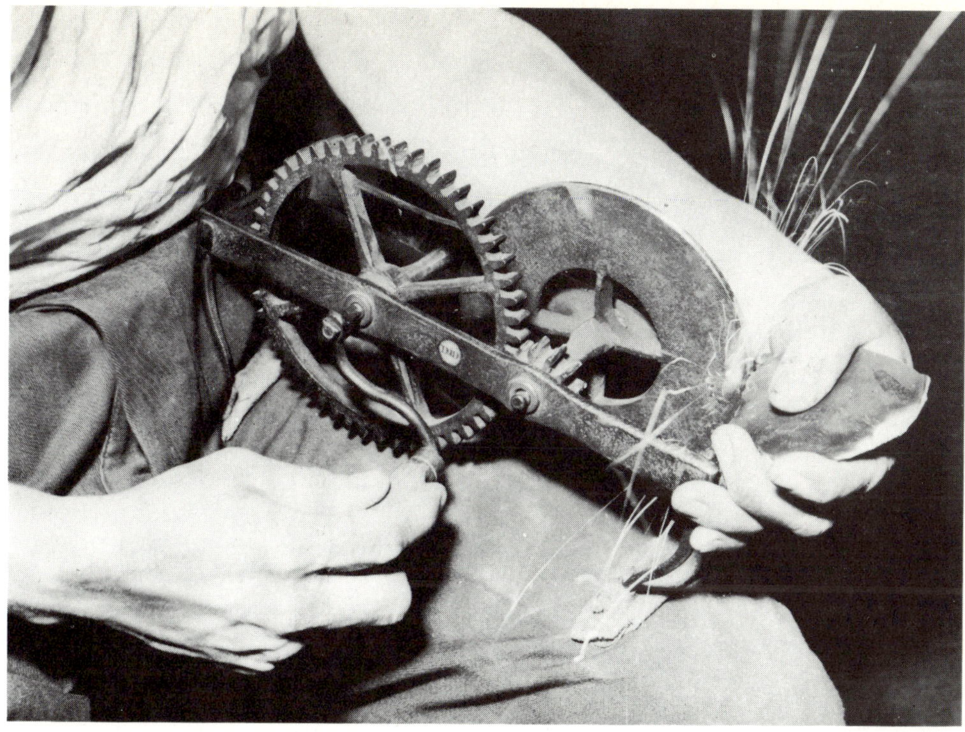

31 This flint mill was one of several new ideas tried in the eighteenth century for making a light to work by in the mines. Quite apart from the risk of explosion, imagine trying to work by the light of a shower of sparks.

adits could not be driven into deep mines. Convection currents produced by a fire at the foot of the shaft drove out the stale air, drawing in fresh air down another shaft. Some of the larger mines on the Tyne and Wear had barriers dividing galleries and gates to divert the flow of air to different parts of the workings, which thereby created the new job of trapping. Blazing fires below ground were not safe or pleasant, and were gradually replaced with steam-driven fans from the 1850s. Many mines still used fires when the Coal Mines Act of 1911 put a halt to the practice.

Many attempts were made to find alternatives to the dangerous candles and tallow dips. Mirrors to reflect sunlight down the shaft were tried in the eighteenth century, and some miners worked by the feeble phosphorescence given off by rotting fish. Tallow dips were preferred by miners, despite the risks. The growth of the industry after 1800 led to more deep pits, with longer galleries. The death rate rose to appalling levels. Six hundred were killed within two years just in pits on the Tyne and Wear. After 92 had died in a single explosion, the Sunderland Society for Preventing Accidents in Coal Mines was formed. This led to the safety lamps designed by George Stephenson and Humphry Davy in

1815. They were certainly safer, and explosions from fire damp became less frequent. However, this encouraged colliery owners to open up even more extensive mines, which caused as many deaths from roof falls, flooding and other accidents.

Another cause of explosion was fine coal dust, which could explode by itself. Though unknown at the time, this had been the cause of many tragic disasters – 145 pitmen were killed at Risca Mine, Newport, in 1860; 361 died at Oaks Colliery, Barnsley, in 1863 and 178 perished in 1867 at Ferndale Colliery, Rhondda. Improvements in air filtration were not enough to prevent an explosion which killed 439 men and boys at Senghenydd, Glamorgan, in 1913.

Quite apart from dangers, there were many difficulties in mining coal. Seams were thin in most of the coalfields. The galleries were kept as low as possible, to save the expense of taking out quantities of useless rock. (Such decisions were taken by managers who never had to work underground.) In Yorkshire, for example, where some of the seams worked in the 1840s were as thin as 25 centimetres, the main galleries ranged from 76 to 56 centimetres high, while those to the face were less. Yet the hewer had to swing a pick lying on his side, while the hurriers dragged wheelless tubs to the tramways. One of the commissioners collecting facts for the 1842 report, S. S. Scriven, visited several mines around Halifax:

> In some of them I have had to creep upon my hands and knees the whole distance, the height being barely twenty inches [50 cm], and then have gone still lower upon my breast and crawled like a turtle to get up to the headings. In others I have been more fortunately hurried on a flat board mounted upon four wheels, or a corve with my head hanging over the back and legs over the front of it, in momentary anticipation of getting scalped by the roof.

The report recommended that young children should not be allowed in mines and that narrow seams should be avoided. The Coal Mines Act of 1842, duly banned women and children under 10 years old from the pits, though they were still employed above ground – women were still sorting coal by hand in Scotland in 1914. Nothing was said about gallery heights. Not all seams were thin, and a high proportion of coal came from seams a metre and more thick. Pit ponies were used to haul the trams in these, living in stables in a worked-out part of the mine. A feature common to all pits was water, which dripped from the roof, seeped through the sides and gathered on the floors. Pitmen were soon soaked to the skin, and had to work in humid conditions in the deeper pits, since temperatures increase with depth.

A lack of concern characterized the colliery owners and managers, who were criticised in the 1842 report:

32 This monument in the churchyard at Silkstone in Yorkshire records the death of 26 people who were drowned when a summer thunderstorm flooded the coal pit. Most were under 12, as may be seen on the two sides.

> One of the most frequent causes of accidents in these mines is the want of superintendence by overlookers or otherwise to see to the security of the machinery for letting down and bringing up the workpeople, the restriction of the number of persons who ascend and descend at a time, the state of the mine as to the quantity of noxious gas in it, the efficiency of the ventilation, the exactness with which the air-door keepers perform their duty, the places into which it is safe or unsafe to go with a naked candle, and the security of the proppings to uphold the roof . . .
>
> There are many mines in which the most ordinary precautions to guard against accidents are neglected, and in which no money seems to be expended with a view to secure the safety, much less the comfort, of the workpeople.

Few industrialists did more than the law required of them at this time and members of the public would doubtless have complained if coal prices had risen to pay for safety measures.

However, it was not just apathy and lack of sympathy that caused mine managers to do so little about working conditions. Draining the mines was a permanent problem. Bucket and windlass, rag pumps and horse gins were used in the eighteenth century. Atmospheric engines were used in the larger mines,

JAMES WATT 1788

SINGLE ENGINE FOR DRAINING MINES

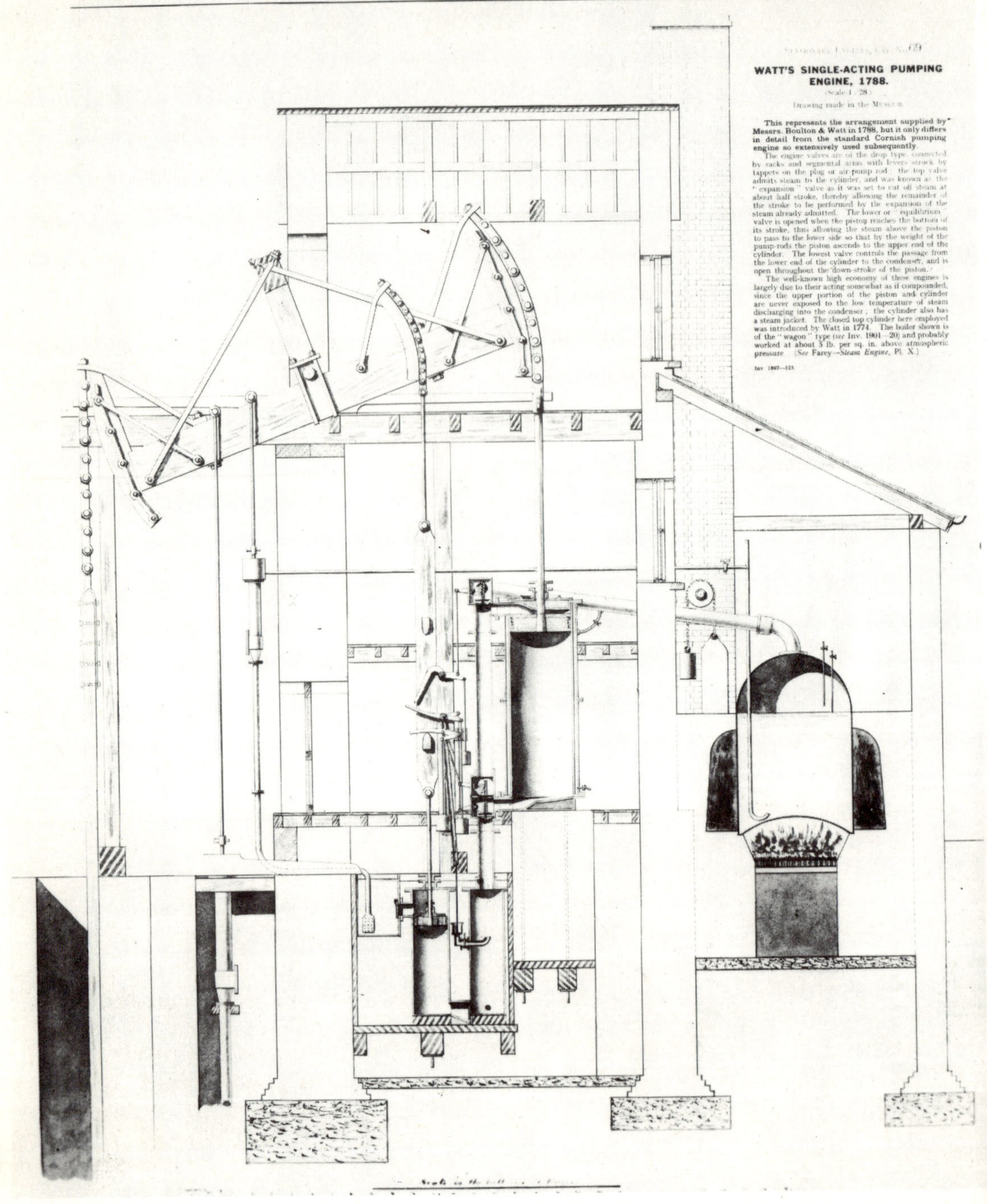

but were too expensive to buy and run for most pits. There were perhaps 200 such engines in use at coal mines in 1800 and a further 30 Boulton and Watt engines, but this was a fraction of the number of pits. Further, the engines had very little power – when 75 men were cut off at Heaton Colliery by an inrush of water from old workings in 1815, it took all available pumps nine months to dry the mine. Winching out men and coal was also done by windlasses and horse gins at many mines in the nineteenth century, though water wheels were used in Scotland and the north of England in the 1760s, and steam engines at some large pits twenty years later.

Miners' unions, particularly the county unions that began to form in the 1840s, were as anxious to persuade employers to improve safety measures as to increase wages. Sometimes they succeeded, sometimes the owners were able to break strikes by bringing in unemployed miners from other areas. Two streets

33 *Opposite* A sectioned drawing of James Watt's improved steam pumping engine. This was much cheaper to work than the Newcomen engines as it used less steam and so less coal.

34 *Below* The economy of working the Watt engine appealed particularly to the tin and copper mine owners in Devon and Cornwall, whose nearest coal suppliers were 200 miles away. Remains of the distinctive houses built for them can be seen in many parts, as here near Helston.

in Burnley are still known as 'Little Cornwall' after unemployed copper miners were taken there to break a strike in the 1870s. Some of the worst conditions had disappeared by 1914, by a blend of union pressure, government regulation and improved technology which allowed improvements to take place, but much coal was still taken out by pick and shovel in cramped and dangerous conditions.

The level of output for any particular mine depended on how easily and how far the coal could be carried to customers. Coal could be taken by wagonway to staithes (waterside quays where vessels were loaded) on the Tyne in 1745 and sold for 5s. [25p] a chaldron; by the time it had been taken by boat via the North Sea and Thames to London, the price was £2.10s. [2.50p]. The Midland coalfields only began to grow when canals were opened – the price of coal in Birmingham fell from 13s. [65p] a ton to 7s.6d. [37½p] during the 1770s for that reason. Those mines that could only distribute coal by road were faced with high costs, shortages of vehicles and restricted markets. However, railways brought almost all the coalfields into a national transport system, leading to the rapid growth in output after 1850.

35 *Opposite page, top* The clothing of a Yorkshire miner in 1812. The steam engine on the hill is being used for winding, which was unusual at this time. The wagonway is the famous Middleton Railway, designed by John Blenkinsop. The engine drew itself along by rack and pinion, and the railway was the first outside north east England.

36 *Left* Very powerful steam engines were developed in the second half of the nineteenth century to lift men and coal from ever increasing depths. The massive cable drum paid out the wire cable to the pully mounted over the shaft. The dial on the left of the engine tenter shows him where the cage is in the shaft.

Chapter 6 Mining for Metals

Much that was said in the previous chapter applies to other kinds of mines and quarries and need not be repeated at length. Long hours were a feature of all kinds of work well into the nineteenth century, and women and children worked in other mines. Many of the dangers found in coal mines were common in others, problems of ventilation applied to all mines and problems of drainage to mines and quarries. But coal mining employed more people over large areas of Britain, so attracting the concern of reformers and leading to reports and legislation. Other kinds of mining grew in the same proportion but never involved the numbers employed working coal, and the differences in method between mining copper and salt were as marked as those between either of them and coal. The workers mining one commodity had little in common with any others, so they had no basis on which to combine to press for improvements.

Iron

Iron was one of the most important ingredients of the Industrial Revolution, increasingly needed for steam engines, textile machinery and machine tools, and later for railways and ships. Iron ore mining therefore increased rapidly, though some was imported as well. The expansion began at the end of the eighteenth century, and accelerated until the 1850s. The ways of making cheap steel that came into use after that depended on imported ore, and mining only expanded again later in the traditional areas of south Staffordshire, Shropshire and the Forest of Dean. It soon extended as new deposits were opened up in Derbyshire, Yorkshire, Scotland and Wales, with isolated mines in Lincolnshire, Devon and elsewhere. The following figures show the trend in output.

Average annual output of iron ore per decade in thousands of tonnes

	Cleveland	Lincs	Leics Northants Rutland Oxon	Cumberland Lancs	Staffs Shrops Worcs	South Wales	Scotland	Total including small areas
1854-63	1,259	17	108	877	1,659	783	1,846	7,618
1864-73	3,527	200	588	1,797	1,708	677	1,711	12,124
1874-83	6,007	794	1,266	2,334	1,925	383	2,415	16,255
1884-93	5,134	1,251	1,547	2,432	1,427	54	1,212	13,478
1894-1903	5,390	1,728	2,265	1,821	945	23	819	13,240
1904-13	5,326	2,138	3,316	1,604	909	29	713	14,756
1914-23	3,515	2,615	3,183	1,261	578	64	301	11,635
1924-33	1,843	2,693	3,523	918	235	107	21	9,350
1934-38	1,708	3,716	5,259	824	140	192	16	11,860

Most of the iron seams worked in the nineteenth century were very thin, from a metre down to as little as 5-7 centimetres. This tended to dictate the methods used. Working conditions were cramped even in the thicker seams, and ironstone was much harder to break than coal. T. Tancred, who had been down many coal mines, wrote in 1841:

> The labour of the ironstone miners is often worse than that of colliers. I have seen them at work in a space of from 22 inches to two feet [55-60 cm] high, where even when seated a man could not keep his neck straight, and to get into the place where he was at work was no easy matter.

37 A team putting in the next props and overhead beam in a section of the Cleveland iron mines. Notice the 12-kilo sledgehammer, and the flames on the acetylene lamps.

38 Many ironstone seams lay almost horizontal and near the surface in the east Midlands. This made it possible to use steam power direct, and large excavators were developed for open-cast mining. These loaded straight into railway wagons. The crew have no protection from the weather.

This was at Whiterigg in west Scotland. Pit ponies could not be used because of the thin seams, so women and older children manhandled corves (large baskets) holding 550 kilos of ore to the shaft. Similar arrangements had to be made in most areas (only in Shropshire were the seams thick enough to use animals), though no women worked underground outside Scotland, and the work was too heavy for young children. Even so, lads of 13 clambered out of Gloucestershire mines with billies (baskets) strapped to their shoulders which held more than 50 kilos. Iron mines were unpleasant to work in, they were wetter and colder than coal mines, and badly ventilated because of the difficulties of forcing the air through such narrow seams. Though there were few explosions, accidents from roof falls were frequent.

Little was done to improve these conditions, and mining methods changed only slowly. Ore was discovered near the surface in Northamptonshire in 1852, and was mined by open-cast working. Steam shovels stripped off the soil, and dug the ore straight from the ground. This method was later used when deposits were found in Lincolnshire. Local needs were met by mining from small open pits in many parts of Britain: Wheal Morley Iron Ore and Clay Works started a small pit near Plymouth, from which they produced iron for the dockyards and clay for bricks and pottery.

Non-ferrous metals

The growth of tin and copper mining during the nineteenth century was startling, and in 1840 Britain was producing three-quarters of the world's output of copper, half the world's lead and 60 per cent of its tin. The copper was mostly smelted in South Wales from ores mined in Devon, Cornwall, Anglesey and Ireland. Small quantities were dug from mines in parts of Scotland, the Orkneys, Merioneth and even the ancient Alderley Edge mine, but most came from the other areas. Mining in Anglesey was restarted after the Great Lode was discovered in 1768. Thomas Williams combined the Parys and Mona mines and, since the deposits were only two metres below the surface, ran them as a single open-cast mine a mile and a half long. It produced 3,000 tonnes of metal a year from 1770-90, but then began to decline.

West Country mines were run quite differently. Each mine was owned by a group of 'adventurers' who had shares in it. They ran the mine on a cost-book system – income and expenditure were added up each year, and any profit left was divided among the adventurers. If there was a loss, they had to pay it off in proportion to the shares they held. The system was simple but meant that mines had no reserves of cash for expensive exploration. Miners were paid in one of two ways: a weekly wage if they were doing tut-work (which included digging new galleries), or a variable sum to those on tribute. Tributers made agreements to work a small section at so much in the pound for a set period, the 'so much' depending on how profitable they expected it to be. The results could be extreme. William Jenkin wrote in 1797:

> Tin Croft does well. It is now indeed very rich. Last month it gained to the Adventurers not less than £3,200 . . . One part of the Mine is so good that the Labourers have just taken a tribute set for one month – and they have engaged to break, rise and cleanse the ores they shall meet with in their limits for having only fourpence [2p] for themselves out of every 20 shillings [£1] worth of ore they shall rise.

In 1804, he wrote:

> I lose no time in informing thee that within the last 6 or 7 days two poor men working in a part of the Mine [Tin Croft] where the Lode was very hard and poor, they suddenly cut into a bunch of rich copper ore, very soft, and consequently easy to break. The poor fellows having by their Contract 12s. [60p] out of the pound for their labour of all the ores they can break, within a certain limit, they scarcely allow themselves sufficient time to eat and sleep since they first discovered the ore.
>
> Their time expires at the end of this month, and if the Lode continues so good as it is now in sight, I expect they will get a hundred pounds each for themselves for two weeks' labour . . .

Base metal mining was so unpredictable that the hope of striking a rich lode made the miners accept low pay and heavy work in a way that coal miners did not. They also worked fixed hours; as early as 1750, Cornish miners worked an eight-hour shift each day except Sunday. The richest veins began to be exhausted in the 1850s, and the cost of working the poorer lodes therefore rose. Far more serious was the development of rich, easily worked mines in the Americas in the 1880s, and in other parts of the world later. These made it cheaper to import copper ore than to mine it in Britain, and many pits closed immediately. Others struggled on by concentrating on the minerals found with the copper, which became merely a by-product. Great Devon Consols turned to producing arsenic, for instance, while miners at Dolcoath in Cornwall searched below the copper and found tin. Most of the attempts to find alternative products were shortlived, and few copper mines were being worked by 1914.

Similar events overtook the tin mines, though at different times. All British tin came from Devon and Cornwall, where there were 75 mines in 1801, 200 in 1838, 230 in 1874 but only 100 in 1900 and just 30 by 1918. Cornwall produced 40 per cent of the world supply of tin in 1865, when the industry was at its peak. The increasing cost of working the Cornish mines and the much cheaper open-cast pits that were developed rapidly in Malaya reduced this high proportion to less than a quarter in 1875 (and to a mere one per cent by 1939).

Most tin mines were worked on the cost-book system, like the copper mines, and mining methods were the same. The only difference was that tin was found

39 The yard of the East Pool copper and tin mine near Redruth, Cornwall, in 1895. The beam engine, pulley wheels on the headstock over the shaft and horse-drawn carts taking ore to the railway could be seen in fewer places at this time as the industry declined.

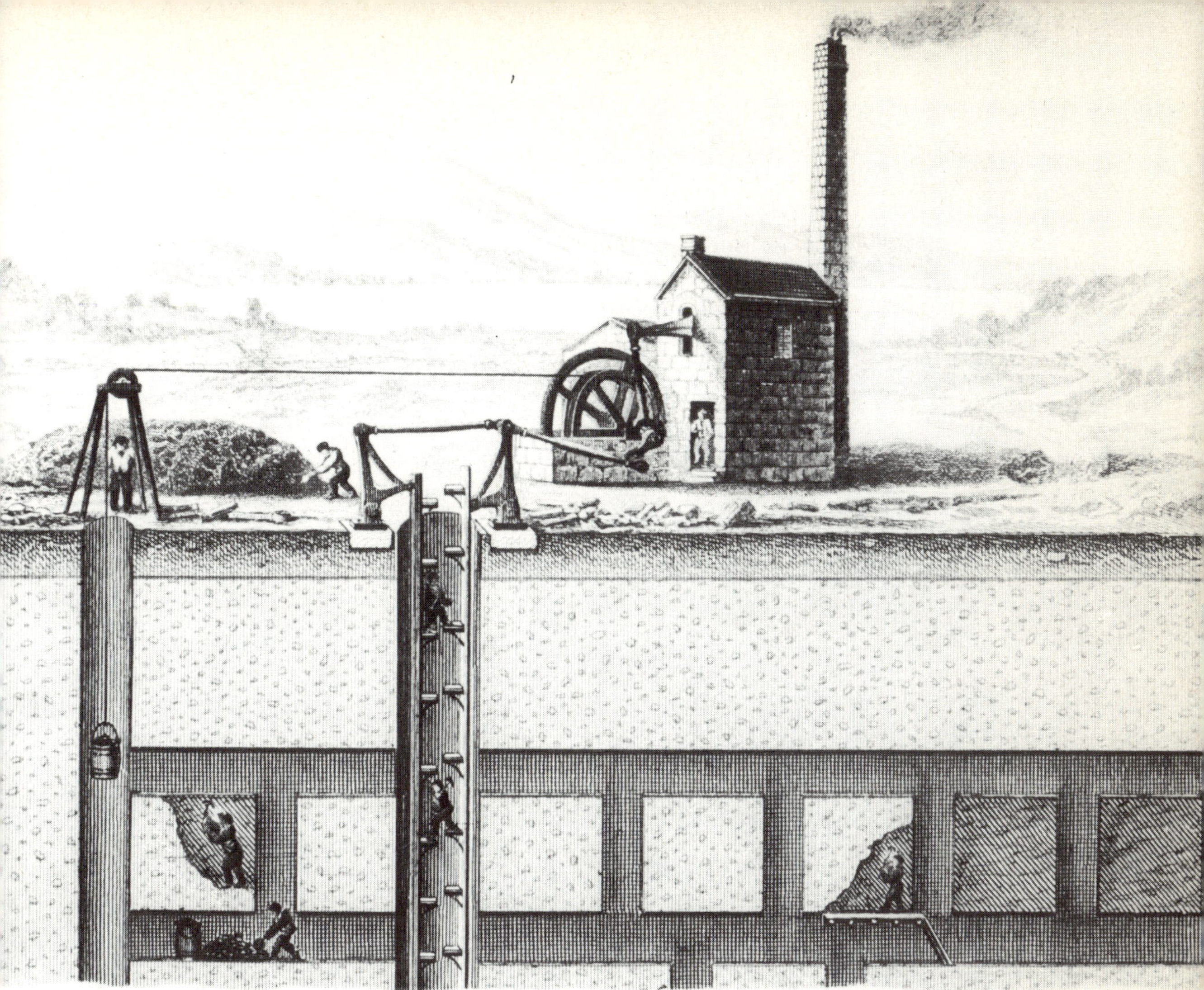

40 A nineteenth-century diagram of a Cornish tin mine which illustrates that mining methods had changed very little over the centuries. The biggest change is the man engine, where the two halves of the 'ladder' rise alternately.

at greater depths than copper, which meant that bigger pumps were needed to keep the mines drained, and that miners had a long climb to the surface at the end of their shift. The climb was up a succession of ladders and could take anything up to two hours. A man-engine was designed in 1842 to lift men vertically up shafts, but only eight were in use 20 years later. This was partly due to lack of interest by the owners, and in many mines, to the fact that the shafts did not go vertically to the lowest level but were crooked and angled as they followed the lode. Dolcoath, the deepest tin mine, had a man-engine down to 240 fathoms, but only ladders for those working on the 362-fathom level, while until 1887 miners at West Wheal Seton had to climb the whole 266 fathoms (486 metres) by ladder. However, by 1900 a few of the largest mines had installed cages where there was enough vertical shaft to make it possible.

Gunpowder was generally used to loosen rock in the mines, but this did not do away with all the handwork. With overhand stoping (working upwards from one level to the next), shot holes a metre long had to be bored with iron boring bars and hammers. Loose powder was rammed in, a straw filled with powder put against it and the hole sealed with clay. A slow match was fixed to the straw and lit. After the explosion, miners shovelled the broken ore down a whinzy (short shaft) to a truck in the level below. Underhand working (digging away the floor) was more often done by moil and hammer. Pneumatic drills began to be used after 1875, and sped up the boring process. But they also led to a tragic increase in the numbers of tin and copper miners dying from phthisis, brought on by the amount of dry dust in the air. The death rate among Cornish miners aged between 25 and 45 was eight to ten times that of coal and iron miners. The mines also lagged behind in other ways. Candles were the most common form of lighting until about 1910, when acetylene cap lamps began to be used. Electricity was hardly used at all, and what the powerful beam engines or water power could not do was done by hand.

Lead mining was more scattered than tin mining, and the most productive areas by the nineteenth century were the Peak District in Derbyshire and the Yorkshire dales, with some also from Wales, Ireland, the Isle of Man, and from reworking spoil tips in the Mendips. The general trend, again due to cheaper supplies from abroad, is shown in the following figures:

Average annual output per decade in tonnes			
	copper ore	tin ore	lead ore
1850s		9,375	92,289
1860s	191,132	13,746	91,276
1870s	77,283	14,389	79,261
1880s		13,898	55,990
1890s		10,777	38,356
1900s		6,567	28,280
1910s		7,289	20,654
1920s		3,405	15,478
1930s		2,873	41,131

The boom years were 1840-70, but by 1875 imports from Spain, Australia and the United States equalled home production, and the industry in Britain rapidly declined. Two thousand miners worked lead in Derbyshire in 1850, but only 200 were left in 1900.

There were considerable differences between the various producing districts. Most of the Derbyshire mines were small concerns employing a dozen or so

41 A windlass was the normal way of lifting ore up the short secondary shafts called winzes. Miners usually climbed ladders, though this one is risking his neck by clinging to the rope. Candles were normal in most mines until the industry collapsed.

men. Simple windlasses lifted the lead from the shafts, and all the work was done by hand. Some of the mines in Swaledale and Arkengarthdale were equally small, though there were more adits than shafts. Further north though, in Weardale, the London Lead Company had invested much capital in larger mines, coal-fired smelting and crushing mills and, at Nenthead, a large village for mineworkers. The same company was active in other parts of the Pennines and in Flintshire, and other companies also conducted large-scale mining. Van Mine in Llanidloes, to take an example, started in 1865, was employing 500 miners by 1870, and added railways, dressing floors and housing. But the fall in prices hit large and small mines alike, and few were making much of a profit by 1900.

Lead mining was hard work as the ore veins are mostly found in limestone. Few children under 13 and no women were employed in the mines, though many were found work breaking the ore on the surface. Most of the accidents in lead mines were caused by falling rocks, as limestone tends to crumble without warning. The lead was not poisonous, though fumes from the smelters were deadly.

42 A large granite quarry at Clee Hill. The squared blocks are setts for road surfacing in towns all over Britain. A stationery engine hauled wagons up the inclines, and horses are still being used.

Other finds

A number of other minerals were prospected and extracted as industry developed and processes became more complex, especially from the middle of the nineteenth century. Far fewer people were employed in this mining than in the coal mines, but the importance of the materials was considerable. Gold was sought to the north of the old Roman workings, particularly around the Mawddach estuary. Some prospectors were lucky, most were not. Some gold was found occasionally in the tin and copper mines but silver was a more certain find. The East Tamar and South Tamar mines struck a vein in the 1840s:

> Vast quantities of silver were brought to London and, on occasions, the whole mail was engaged to carry it. The establishment in London consisted of seven directors with large salaries . . . The offices in Bishopsgate Street were like a palace and an usher with a gold stick stood at the door.

It is perhaps not surprising that the company soon failed. Most mines that found silver treated it as a stroke of luck, and carried on with what they were doing, though Wheal Betsy in Cornwall mined silver with lead from 1806 until 1877. Fluorspar was obtained with lead in some parts of the Peak District and Yorkshire dales, and was used in the new steel furnaces at the end of the nineteenth century. Among other metals, wolfram (and also plumbago) was mined in the Lake District, wolfram and molybdenite in Cornwall, barytes in Cornwall, Shropshire and the Lake District and zinc in Wales, Shropshire, the Pennines and Isle of Man. Few countries had such natural advantages in the variety of minerals available.

The same applied to types of stone, though less and less was quarried for domestic buildings. Many stately homes were rebuilt along classical lines in the eighteenth century, and local stone was used for houses wherever it was satisfactory. The new local authorities created in the nineteenth century had to have their ornamental town halls and offices, as did the growing number of government departments. All used stone. The popularity of brick was increasing even at that time, and leaped ahead with the abolition of taxes on bricks in 1850. This and new machines which could handle the stiffer clays found in the Oxford Clay Belt led to nearly all the new buildings in the industrial towns being brick, especially from the 1880s. One works alone (Fletton, near Peterborough) produced 156,000 bricks a day in 1889. The new houses were roofed in slate, which was produced in vast quantities from quarries in North Wales as soon as canals made it possible to transport it cheaply to the Midlands. The Penrhyn quarries alone produced 12,000 tonnes of slates in 1792. The tax on slates was ended in 1831 which, combined with the developing railways, made slate the obvious roofing material in all the rapidly expanding urban areas.

Stone was used in considerable quantities on the roads. Those turnpike trusts that employed engineers were advised to lay foundations half a metre thick, which added up to about three tonnes of stone for every metre of new road. Even the parish roads had to be repaired with stone, which was brought to the roadside in large pieces and broken up by men sent from the parish workhouse. This prompted G. Walker to write in 1814:

> In times like these, when machinery is applied with profit and advantage to almost every purpose of agriculture and trade, it must be a matter of great surprise that no machines have yet been invented and used for breaking stone for the road.

This comment could be applied equally well to nearly all forms of mining and quarrying in this period, for the use of machinery and power lagged far behind the methods of manufacturing industry, for which the minerals were obtained.

Chapter 7 Hard Times

The First World War (1914-18) provided some of the busiest years for many of the mining industries, delaying the difficulties that threatened to prove overwhelming. The war was the first conflict to see a systematic blockade, made possible by the German fleet of submarines. The consequent shortage of shipping and other transport encouraged the maximum use of iron ore deposits, and the many uses of coal kept coal mines at full stretch. Coal was so important that the government took over the mines in 1916 in order to guarantee that sufficient was mined.

The unnatural needs of the war years, however, hid the real difficulties facing most of the mining activities. Lead, tin and copper mining was no longer profitable in the face of cheaper supplies from abroad. Iron mining was profitable but depended utterly on the fortunes of the steel industry which was less

efficient than competitors in Germany and the United States. The coal mines that had been the mainstay of the Industrial Revolution and its aftermath were facing many problems, chief of which were very bad industrial relations, the rising cost of mining and the slowly shrinking demand for coal. Even clay mining had to face some decline as reinforced concrete heralded the wider use of new building materials. Only such materials as ball and china clay, salt and the raw materials of the chemical industry could look for reasonably normal business.

All these difficulties would have fallen on the mining industry in any case, due to consumption of the mineral deposits, changes in the need for minerals in this country and abroad, and a persistent refusal to modernize management methods to keep up with changing economic trends. The changes were made the more devastating by the artificial boom in the demand for mined materials during and immediately after the war, and the depression that began to overtake the economy from 1920. In terms of numbers, the coal mines were the worst hit. Thousands of miners lost their jobs as the market for coal shrank, and many did not regain them when trade revived somewhat towards the end of the 1930s. The effect on some other mining activities was more final. Lead mining was virtually extinct by 1945, and tin mining seemed about to vanish after a continuous history of 4,000 years. Stone quarrying was in decay, in the face of rising costs and fading fashion, and many quarries were abandoned. Among them were some of the largest of the Welsh slate quarries, beaten by the widespread use of cheaper tiles.

Coal

The key role of coal in the nineteenth century makes it the obvious starting point for a study of the decline of mining. The highest peak of production had been reached in 1913, when 283 million tonnes were mined. Most of this had been obtained by outdated methods: 92 per cent had been cut by hand-held picks, almost entirely by the light of safety lamps, and most of it had been loaded by hand into tubs to be moved by men and pit ponies. Although half of the pits had electricity, only a few used it for any more than lighting the area around the bottom of the shaft and the main headings. The general trend for the past 30 years had seen output per man falling because the easier seams had been exhausted, while narrow, faulted seams were increasingly being worked to keep up with demand. Falling productivity brought rising costs, which meant that those colliery owners who had not saved money when trade was good were in no position to buy machinery, install electricity or adopt any of the other

43 Striking miners fill the colliery yard, while police try to clear a path for the managers and secretaries. Similar scenes could be seen at most pits in the 1920s.

44 Long-wall working was used almost everywhere by the 1920s. It was still mostly pick and shovel work — the lack of machinery is striking.

improvements coming onto the market. The majority of owners found themselves trapped between rising costs and a falling market for their coal. Any chance of using cost-reducing methods was remote, yet that was the only course that could save them from disaster.

The demands of war provided a temporary refuge for the colliery owners, for coal was needed to make gas and electricity, to produce steel and countless other vital materials, as well as in the home. Nearly all the export trade, though, vanished, as this had mainly been to countries in central Europe. Sales within Britain reached slightly higher figures than ever, though the loss of overseas customers meant that the total output, at about 250 million tonnes a year, was less than the 1913 figure. The government used its emergency powers to take over the mines in 1916 to ensure that the various types of coal were produced in adequate quantities. The next few years provided a clear example of the problems created by governments when they try their hand at organizing industry. The mines were run for the rest of the war on a hand to mouth basis. Little maintenance was done beyond what was needed for safety, and much surface equipment was sent to France for use in the supply lines. It was hardly a coherent policy, though admittedly the government was no less successful at producing coal than the mine managers.

The end of the war brought to the fore the question of what should be done with the mines. The Sankey Commission was appointed to examine the problems of the industry, and made its report in 1919. It recommended that the government should nationalize the coal mines fully by buying out the mine and mineral owners. It also called for higher wages and a seven-hour working day (though the time taken by a miner to reach the face would still be additional and unpaid). The government ignored the advice and handed the mines back to their owners in the run-down condition that they were now in. Few mine owners had any reserves of money, either to pay higher wages or to install machinery that could boost output per man. Miners and mine owners were slow to realize that the old days when customers fell over themselves to buy coal were over. A strike by the National Union of Miners in 1921 for higher pay did nothing except leave a legacy of bitterness, and relations had been bad enough before. The government meant to help by giving subsidies to the owners, which were used both to raise wages and buy machinery, but when they tired

45 The pit-head baths at Ellington Colliery, Northumberland. Each miner had his own hook for his surface clothes when he went down the pit, and put his pit clothes there after a bath.

of that approach and withdrew the subsidies, it brought on the disaster of a futile six months' strike in 1926.

The root cause of these troubles was that the industry was contracting. The export trade was gone, as former customers had developed their own resources during the war. The home market was beginning to decline in the face of competition from oil, and also because many of the largest consumers of coal were finding more economical ways of using it. To take one example, it needed 1.4 kilos of coal to generate one unit of electricity in 1913, but only 0.46 kilos by 1937. Similar economies were made in such major coal-consuming processes as the production of gas and iron smelting, and they had a cumulative effect in reducing the sales of coal. The result was growing unemployment in the industry. The problem was worse in some areas than others – South Wales and the north-east were especially badly hit since they had mined nearly all the export coal. Trade in the other mining areas varied with the fortunes of the basic industries in each region – Lancashire mines suffered with the cotton industry, while the Midlands were relatively unaffected.

A Coal Mines Act was passed in 1930. A Coal Mines Reorganization Commission was created, with powers to enforce amalgamations among mine owners. Only a handful of voluntary mergers were in fact made, but a more effective part of the Commission's work was the setting of quotas and minimum prices, which helped to make the industry a little more profitable. The next ten years saw a progressive decline in the number of men employed in the industry – numbers fell from 1¼ million in 1913 to ¾ million by 1938. Their productivity was steadily increased by the hasty installation of electricity, cutting machines, conveyor belts and motor haulage below ground. Even so, British mines lagged far behind those in the United States.

% mechanically cut		
	UK.	US.
1913	8	
1920		60
1921	14	
1929		78
1930	31	
1938	59	

Similarly, electric winding was widely installed in France, Germany and the United States between 1905 and 1914, but was still only in use in 182 British pits in 1924. These figures underlie the backlog of modernization needed in the mines, the great difficulty of adapting machinery to the thin and faulted

46 The miners organized their unions by counties in the nineteenth century, and the Durham union stayed out of the Miners Federation when it was formed in 1882. Durham miners ran a nursing home for injured miners, and particularly for those suffering from silicosis.

seams that made up the bulk of those being worked in the 1920s, the difficulty of finding money to invest in such improvements and the upheavals that all those connected with the industry had to face.

The quota and minimum price arrangements made in the 1930s stimulated the move towards mechanization. They also led to the general use of an updated version of long-wall mining, known as cyclic mining. Three shifts of miners kept up a continuous pattern of work along these lines: the first shift undercut the coal and brought it down, the second cleared it away and the third built up the stone wall, extended the roadway and moved the conveyor forward. A new machine that both cut and loaded the coal on to the conveyor came into wider use during the Second World War (1939-45) and meant that two of the three shifts were producing coal. The war years again brought a revived demand for coal, and the mines were once more taken over by the government (in addition to the unworked deposits which had been nationalized in 1938). Some of the depression had lifted from the industry but there was a residue of bitterness stored for the future. Most mining communities had been scarred by unemployment, and by the breaking of ties as men made redundant moved away in search of work.

Metals

Iron ore mining was directly connected to the state of the iron and steel industry, the only customers for ore. Exports of iron and steel goods fell sharply during the depression years, and cheap imported goods made the home demand for steel less than it had been. These events had a direct effect on iron ore mines – some were closed and the number of miners fell by nearly half between 1923 and 1939, to only 11,000. The government started to help the steel industry in 1932 by taxing imports, and the revival of trade then and even more when rearmament began in 1937 led to the opening of new mines and considerable mechanization. About 60 open-cast pits were in production in 1933, mostly in Lincolnshire, and steam shovels and larger machines were used to strip off up to 30 metres of overburden as well as ore. One illustration of the new confidence was Stewart & Lloyds new £3-million tube factory at Corby, Northamptonshire, which used half a million tonnes a year by 1939.

The non-ferrous mines had no such encouragement. Copper, gold and silver mining had ceased, and more tin was obtained by dredging the sand washed

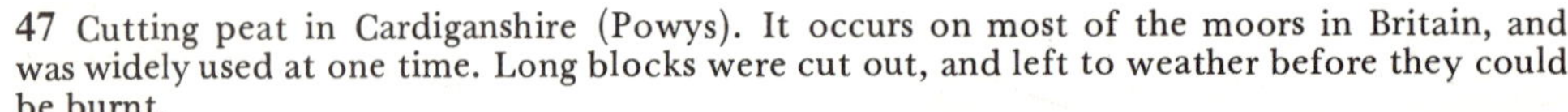

47 Cutting peat in Cardiganshire (Powys). It occurs on most of the moors in Britain, and was widely used at one time. Long blocks were cut out, and left to weather before they could be burnt.

48 Open-cast iron mining was extended to most mining areas in the 1930s, and the size of the mechanical shovels increased.

49 Miners engaged in overhead stoping at the King Edward tin mine, Cornwall. The one on the left is cutting away small pockets of ore, while the pair on the right drive a shot hole. The unpredictable course of a tin lode makes twentieth-century methods resemble fig. 13, but there are some differences.

from mines around Camborne into the River Fal than by deep mining. Lead mining was slightly more active. Halkyn mine in Flintshire and Mill Close in Derbyshire were reopened in the 1920s, and between them produced 68,000 tonnes of lead concentrates in 1934. A few other mines produced lesser quantities, but prices for lead were low and not enough to make British mines pay. Halkyn closed for good in 1941 and Mill Close in 1944, both unable to make enough money to pay for adequate drainage. An official committee produced the Betterton Report in 1920, which forecast that these mining industries would not revive without government help. Other committees in later years said the same, but neither subsidy nor tax relief was forthcoming – governments could afford to ignore the handful of voters employed in non-ferrous mines. The Second World War caught Britain dependent on imported supplies. There was a frantic search among old spoil tips but only one at Nenthead, County Durham, was worked for zinc.

50 A substantial stockpile of bricks in the east Midlands, with the brickworks behind to the right. The clay came from open pits nearby, the coal from further away.

51 *Above left* Much roofing slate was dug from open pits and quarries as in Cornwall and north Wales. The bigger blocks came from deeper mines in Wales and the Lake District. It was used for doorsteps, windowsills, tombstones and much more.

52 *Above right* Beach sand was seldom used for building because of the impurities in it. In some areas, though, it contained so much crushed shell as to be a good fertilizer. The wagonway across the beach at Bude, Cornwall, took sand to canal barges for farms inland.

53 *Right* This advertisement of 1928 indicates a few of the many kinds of stone that were still in common use at the time. This firm was one of the many involved in adapting roads to suit cars, and grew by absorbing other quarries.

OUTPUT EXCEEDS 500,000 TONS PER ANNUM.

The BRYANT & LANGFORD
GROUP of QUARRIES

EMBRACING

Tresarrett Quarry Co., Ltd.
Black Rock Quarries, Ltd.
Winterbourne Quarries, Ltd.
Bryant and Langford (Malvern), Ltd.

Penderyn Limestone Quarries (Hirwain), Ltd.
Stoneycombe Lime and Stone Co., Ltd.
Keinton Limestone.
Dorset Quarry Co., Ltd.

MACADAM, TARRED MACADAM AND CHIPPINGS.

STONE and GRANITE

For all Building, Engineering and Monumental Purposes.

Contracts for laying Tarred Macadam or Chippings, with or without maintenance, undertaken in any part of the country.

BRYANT PRODUCTS.

Malvern Granite Macadam and Chippings. All grades, either dry or tarred.
Limestone Macadam and Chippings. All grades, either dry or tarred.
Black Rock Macadam and Chippings. Either dry or tarred to any required gauge.
Langton Limestone Macadam and Chippings. Dry or tarred.
Cornish Elvin (Granite) Macadam and Chippings. All sizes.
Bristol Pennant Stone. For Railway, Dock and any Engineering requirements. Kerb and Channel. Rough Rubble for Bottoming to Roads. Walling Stone, Shoddies, Steps, and all classes of Building Work.
Keinton Limestone. For Kerb, Channel, Crazy Paving, Dry Walling, etc.
Ground Limestone. For Filling and Agricultural Purposes.
Grey Cornish Granite.
Forest of Dean Stone.
Bristol Pennant Stone.
Portland Stone.
Bath Stone.
For Building, Engineering, and Monumental purposes of all descriptions, either in the Rough, or Sawn, or dressed, or Fixed and Cleaned Down Complete.

BRYANT Products are renowned for their high grade uniform quality and economical prices. Practically inexhaustible supplies, coupled with the promptitude with which we despatch our orders, enables us to deliver any quantities at very short notice.

Our work is so organized that we can give regular daily despatch of any quantity ordered to accord with your own arrangements for haulage and use, if it is more convenient to you.

Let us quote you for supplying and delivering any of the materials shown in our list.

WALTER BRYANT

QUARRY OWNER and STONE MERCHANT. ::

Western Chambers, 32 Nicholas St., BRISTOL

Telephone: Bristol 5250 / 5251 Telegrams: "Macadamize, Bristol."

54 A ball clay mine in the 1930s. The clay was being raised from 50 metres, but was still cut in the distinctive blocks. No heavy pit gear could be used because of the softness of the ground.

Non-metals

Hardly any stone was quarried for building in the first half of the twentieth century, but increasing quantities were needed for road maintenance. The sudden arrival of motor vehicles on the roads made it necessary to resurface them with tarmac. This undramatic use of stone, like that of chippings for concrete, was the main reason for a 66 per cent increase in the numbers employed in stone quarrying between 1923 and 1939. The increase in the numbers employed in extracting clays, sands and chalk was almost as great, keeping pace with the construction of new brick houses, the production of glass and tableware, and so forth.

A wide variety of raw materials for chemicals was produced in these years. The quantities were often slight, and so the numbers of men employed were never great, but production was often enough to supply all that was needed. About 2,000 tonnes of white arsenic were mined annually in Devon and Cornwall, for use in insecticides, dyes and paint. Production declined in the 1930s as safer alternatives were found. Mines in Derbyshire and County Durham produced an average of 40,000 tonnes of fluorspar a year, which was used by the steel industry as a flux, while barytes was obtained from mines in many districts. Anhydrite was mined on a small scale in Durham, Cumbria and Yorkshire during the First World War, and large-scale production started on the substantial deposits at Billingham-on-Tees in 1929. It was used to make sulphuric acid. It would be tedious to list all the materials mined for producing chemicals, but just as Britain has been blessed with most of the metal ores needed in the eighteenth century, so now it could meet the basic needs of chemical manufacturers. Only oil seemed an impossibility.

Chapter 8 Today and Tomorrow

The dispiriting years between 1914 and 1945 seemed endless to those who lived through them – a war to end wars, talk of a land fit for heroes ending in the wasted years of the depression and yet another war. Yet in reality it was only 30 years. Understandably, many people wanted to prevent a similar sequence of events following on the end of the Second World War, and some of the planning led to far-reaching changes. The changes were most drastic in the coal mines, since they made up the largest single mining industry and had been declining for some years. The Labour Party had made it plain between the wars that they would nationalize the mines as a matter of urgency as soon as they had the power, and they lost no time in doing so when they were elected in 1945. The National Coal Board became responsible in 1947 for all but a handful of small mines – the miners of the Forest of Dean retained their freedom. The Socialists hoped that a nationalized industry would treat its workers far better than the private mine owners had done, would be able to find the money to modernize the pits and, with the profits going to the nation, would reduce taxes.

Some of these hopes were fulfilled, even though it took time. The National Coal Board was able to apply to Parliament for loans and grants totalling millions of pounds, which it used to provide pit-head baths and to install adequate lighting, safety helmets and other necessary equipment. It also set about a long-term programme to modernize the industry. Pits that were losing money, like most of the Somerset mines which lost £1.50 on every tonne mined, were closed down, and the miners taken on at pits in other coalfields. New pits were opened and others were re-equipped with the latest machinery. Some of the new mines were open-cast, though annual output from these seldom exceeded seven million tonnes.

Cyclic mining became continuous mining in some of the largest mines in the 1950s when cutter-loaders and hydraulic pit-props made it unnecessary to have a whole shift employed moving machines and shoring up the roof. This boosted output to six-and-a-half tonnes per man per shift. Far more advanced was a system called ROLF (Remotely Operated Long-wall Faces), which was tried out at Bevercotes Colliery in Nottinghamshire in 1963. This used continuous cutting machinery remotely controlled with the aid of closed-circuit television. ROLF produced 23 tonnes per manshift when it was used under normal conditions.

Another aspect of the modernization programme was the search for new coalfields. Substantial reserves were found in Oxfordshire and Nottinghamshire,

55 The console in Ormonde Colliery that controlled ROLF equipment. The single miner was the only one working in the gallery while the coal was being cut.

and 500 million tonnes were found in such thick seams around Selby, Yorkshire, that immediate plans were made to start mining it.

However, nationalization was unable to meet all expectations for it was, after all, only a change of management. The profits that had been looked for to share among the people became colossal debts to add to taxation, with few profitable years in the first 25 as a nationalized industry. The coal mines were only one of several major industries nationalized in the 1940s. There was a serious shortage of people who could think on the scale required to manage such concerns, which meant that mistakes were made and there was a permanent wastage of materials and money. There was no sudden end to the bad labour relations either, for the NCB were still the bosses who decided rates of pay and similar matters. Yet the Board had bosses too, in the shape of Parliament, and successive governments intervened in the running of the mines to an extent that made it hard to tell who had the last word. Inevitably it became normal to think, as in everything to do with spending so-called public money, that expense was no object since the government had the power to demand higher taxes from the people to pay off the debts.

The Board's greatest difficulties stemmed from the problems of finding customers for the coal. The trend towards greater economy in the use of coal continued after 1945, but was partly offset by continuous increases in the demand for power. New electricity generating stations, even efficient ones, represented valuable customers for coal. It was government policy until about 1965 to encourage the use of coal in order to keep production at about 200 million tonnes a year. But the advantages of cheap oil, which was much easier to handle, began to seem more appealing and the government's interest in atomic power forced the coal industry into a painful decline – output was less

56 It has proved pointless to draw a line in this book between mining and quarrying, since many minerals have been obtained by both methods. So it seems sense also to include the bringing ashore of natural gas. This production platform is in the Forties Field in the North Sea.

than 150 million tonnes in 1970. A number of mines were shut down in all areas, including Mosley Common Colliery which had been a Lancashire show-piece.

The sudden increases in crude oil prices from 1974 showed that the rundown of coal mining had been premature, and proved yet again the folly of relying on imported basic materials. The National Union of Miners were quick to use their scarcity value to force up wages, for the NCB could not open mines as quickly as they had closed them. The first 30 years of nationalization were an unhappy period for the coal mines. Miners found that nationalized bosses were no better at retaining full employment than private ones had been and the country saw little return on the millions of pounds poured into the mines. Unfortunately, the mere nationalization of isolated industries could not work without thinking out the nation's needs on a wider scale. The means for doing this evidently did not exist, so nationalization only passed control of the mines to politicians who could not be expected to make a success of the business. People who could think out long-term needs were desperately needed.

Oil and gas

Some gas was raised from wells in Yorkshire in the 1960s. The race to find reserves of natural gas under the North Sea began when the Dutch found some there in 1959. Many companies, British and foreign, spent millions of pounds finding and bringing gas into Britain, the first in 1967. There was so much that the Gas Board immediately began to convert every gas burning appliance so that the new gas could be used instead of gas made from coal or imported oil. The search for oil yielded success and the first was brought ashore from the North Sea in 1975.

57 *Left* The Forties Field also produced the first North Sea oil, which was sent by pipeline to the refinery at Grangemouth in November 1975.

58 *Right* In contrast to the very costly equipment used to bring gas and oil ashore was the primitive equipment used in the St David's gold mine, in the Clogau area of Wales. More than 1,000 kilos of gold had been produced in the area between 1900 and 1905, but much less in the 1970s.

Metals

Iron ore mining was as always directly related to the fortunes of the steel industry, and these varied considerably after 1945 due to nationalization, the change from producing war goods to peace goods and the stop-go policies of successive governments. Output averaged four million tonnes a year but varied up to 20 per cent a year either way. Most of the iron ore came from existing mines.

Developments in the other metals were more varied. The last lead mine in England closed in 1962. This was Greenside mine in Glencoynedale in the Lake District, which had been producing small quantities of lead and other minerals. By contrast, the first new mine to be started in Eire for a century began at Tynagh, County Galway, in 1965. It produced lead and zinc, and was joined by another at Silvermines in 1968. Eire was producing 60,000 tonnes of lead a year by 1970. Copper mining was defunct, but a small copper and silver mine was opened in Eire in 1967.

Several attempts were made to revive tin mining, and surveys were made in the Isle of Man and other areas. Few of these led to any permanent mining, except in Cornwall. Mines at Levant and Boscaswell Downs were reopened in 1969, and a new mine, Wheal Jane, was started near Truro in 1971. Tin production exceeded 2,000 tonnes by 1974 (when it was selling at nearly

£4,000 per tonne). Tin has now been mined in Cornwall for over 4,000 years. Attempts to restart gold mining in Wales illustrate a problem that increasingly hampered mining engineers. In 1972, Rio Tinto Zinc proposed to start dredging the Mawddach estuary around Barmouth in order to recover gold that had been washed down from the mines active on the tributary rivers in the nineteenth century. There was an immediate outcry from conservationists, who were concerned about the effects on wildlife in the area. The firm abandoned its plans the following year, though limited gold mining was resumed at Dolau Cothi to pay for archaeological research.

Chemicals

One of the few mining activities to experience continuous expansion at this time was salt production. Apart from its use in cooking and preserving food, it was also needed as a raw material in industry. Production between 1962 and 1971 rose from 6.1 million tonnes to 9.2 million tonnes. Most of the salt came from the Northwich-Middlewich-Sandbach area of Cheshire, and other mines were at Droitwich, Middlesbrough and Fleetwood. Except for a single mine where rock salt was cut, the salt was obtained by forcing water underground, pumping out the brine formed and then drying the salt.

Several other minerals were mined for use in industry. A large new fluorspar mine was opened near Eyam in Derbyshire in 1964 to provide flux for the steel manufacturers, and ICI and the British Steel Corporation began to work some of the old lead and iron mines in the North Pennines in 1972, also for fluorspar. Gypsum continued to be mined in Sussex, and production was increased by the opening of a new mine in 1963. Controversy was again aroused by plans to mine a million tonnes of potash a year from sites near Staithes and Whitby. They lay within the Yorkshire National Park, and there were fears for the scars that mining would leave on the landscape. Mining was started, though, under landscaping conditions.

Those employed in extracting sands and clays had many changes to face after 1945. Wartime air raids had damaged many buildings, which meant that there was a period of rebuilding which lasted well into the 1950s. This caused the brick-makers to be as busy as they had ever been, particularly those in the Peterborough area who had the natural advantages of a clay that was both near the surface and cheap to fire. Bricks began to lose favour at the end of the 1940s, due to the cost of transporting and laying them and also because of a reaction against the monotonous sameness of brick buildings. Ready-mixed concrete was easier to handle, and offered the possibility of different shapes and appearances. This meant more work in the gravel pits at the expense of some of the larger clay pits. Some small local brickworks, however, had more work because their products were unique to the area, even though they were more expensive.

Sands were also extracted for making glass. The demand for all kinds of glass increased steadily after the war and many new sand pits were opened, particularly in eastern England. Sand and gravel were easily extracted from open-cast pits by excavating machinery; the problems, environmental ones, came when the pit was worked out and abandoned. The demand for pottery clays also grew, to make goods ranging from porcelain to drainpipes. China and ball clay producers equipped themselves with modern machinery to enable them not only to meet the requirements of British firms but to develop a substantial export trade as well.

Britain is almost unique in the range of minerals that are available in the ground. The discovery and use of these made it possible for Britain to have several flourishing industries even in Roman times, and to develop these further in later centuries. The minerals have been the raw materials or sources of fuel that have allowed industry to develop. In recent years, many of the extraction industries have declined or disappeared due to exhaustion of the deposits, shifts in the demand for some minerals, or the straitjacket of short-term gain on the part of governments. It is doubtful whether an industrial country can afford to rely on imported materials indefinitely, when it could be devising new uses for the raw materials it already has.

59 Ball clay mining in the 1970s, by large excavators and dumper trucks in open-cast pits. The dark bands are lignite. The extraction of sand and clay were among few examples of growth in mining industries at this time.

Further Reading

Among **shorter books on various aspects of mining** are B. W. Smith, *Sixty Centuries of Copper* (Hutchinson 1965), J. Temple, *Mining, an International History* (Benn 1972), R. H. Bird, *Britain's Old Metal Mines* (Bradford Barton 1974), J. R. Hamilton and J. F. Lawrence, *Men and Mining on the Quantocks* (Town and Country Press 1970), A. K. Hamilton Jenkin, *The Cornish Miner* (David & Charles 1972), and, in the Then and There series, *An Iron and Coal Cummunity in the Industrial Revolution* (Longmans 1965). Methods and machines can be found in T. K. Derry and T. I. Williams, *A Short History of Technology* (Oxford 1960), and the place of mining within the context of other industries in H. Bodey, *Twenty Centuries of British Industry* (David & Charles 1975).

There are many detailed books on **particular industries**, even about individual mines, and about the way in which the materials were used. Among the more recent on coal mining are T. S. Ashton and J. Sykes, *The Coal Industry of the Eighteenth Century* (Manchester University Press) and A. R. Griffin, *Coalmining* (Longman 1971). Metal mining is described in T. D. Ford and J. H. Rieuwerts, *Lead Mining in the Peak District* (Peak Park Planning Board 1968), J. B. Richardson, *Metal Mining* (Longman 1975), D. B. Barton, *A History of Copper Mining in Cornwall and Devon* (Barton 1968), C. Noall, *Botallack* and *Levant: the Mine beneath the Sea* (Barton 1972, 1971). Clay and stone are best seen through the eyes of A. Clifton-Taylor, *The Pattern of English Building* (Batsford 1962). One book about sixteenth-century methods in Germany must also be mentioned, Agricola's *De Re Metallica* (Dover Publications), as it is the only book with so much detail from this period. All these books are long and detailed, so it would be better to dip into one or two than set out to read them all.

The **physical remains** of mining are to be found in most counties. However, it is highly dangerous to enter any abandoned shafts or adits, and just as dangerous to walk alone on those moorland areas where open mine shafts are frequent. There are plenty of places where much can be learnt of mining work in perfect safety. Most spoil tips and surface workings can be seen from a safe distance, for example. The canal and part of the sand wagonway still stand in Bude; the remains of wagonways and smelting works can be seen in Swaledale and of the Roman aqueducts at Dolau Cothi.

Among **museums** with special interests in mining are the Lound Hall Mining Museum, Retford, Nottinghamshire (coal), the museum of the Department of Engineering, Newcastle upon Tyne (coal), and the museum at the Lizard in Cornwall (tin and copper). A water beam engine can be seen at Wanlockhead, Dumfriesshire, and steam pumping engines in the Science Museum, London, while the National Trust preserve the one at East Pool in Cornwall and four others in the county. The NCB have an old beam engine at Elsecar Colliery near Barnsley but it is seldom open to the public. A horse gin has been moved to Shibden Hall Museum in Halifax. Various styles of brick nogging have been built in the Bridewell Museum, Norwich. Visitors are welcomed at the Delabole slate quarry in Cornwall, and can travel on a wagonway through the old slate mine workings at Llechwydd near Blaenau Ffestiniog, Gwynedd. A museum of the china clay industry is in the process of being created at Wheal Martin near St Austell in Cornwall. Further sites are also listed in H. Bodey, *Discovering Industrial Archaeology and History* (Shire 1975) and in the gazeteers of the regional industrial archaeology series published by David & Charles.

Among **novels** with a mining background are Alexander Cordell, *Rape of the Fair Country,* Richard Llewellyn, *How Green was my Valley*, and Winston Graham's Poldark novels: *Ross Poldark*, *Demelza*, *Jeremy Poldark, Warleggon, The Black Moon* and *The Four Swans.*

Glossary

Adit A tunnel driven into a hillside to gain access to a seam. It is driven at a slight slope so as to be self-draining. Some adits were dug solely for drainage.

Base Metals Those other than the precious metals of silver and gold.

Battery Making wire by battering metal through a hole in a die.

Brass An alloy of copper and tin or zinc.

Charcoal Wood that has been partly burnt in a reduced air supply.

Coke Coal that has been partly burned in a reduced air supply.

Chaldron A unit of measurement by volume but one which had regional meanings. A chaldron of coal in the eighteenth century was 53 cwt in Newcastle but only 25½ cwt in London.

Day hole, Dene hole A shallow pit or small adit mine, sufficient to provide the needs of a family day by day.

Drift A sloping entrance to a mine as opposed to a shaft. Used at coal mines in recent years to allow the continuous removal of coal by conveyor belt.

Edge tools Those with a sharpened cutting edge.

Fault The movement of one mass of rock in relation to another, causing seams to separate.

Fire-setting Shattering rock by alternately heating it and cooling it rapidly with cold water.

Gallery A passage radiating from a shaft where a seam is being worked. They are seldom horizontal.

Gin A horse powered windlass, used to lift rock from mines. Sometimes called a whim or win-gin.

Hewer The man who cuts the coal from the seam.

Lode A seam or vein of ore, a word usually used in connection with tin and copper mines.

Moil A wedge hammered in to split a rock.

Non-ferrous Other than iron (and steel).

Nationalize For the government to buy out and, ideally, run for the benefit of the nation.

Open-cast A pit open to the sky, making it easier to use large machinery and avoiding the problems of ventilation, roof falls and the sinking of shafts.

Palaeolithic The Old Stone Age, very approximately between 1 million years B.C. and 10,000 B.C.

Pewter An alloy of tin and lead.

Prospecting Looking for and testing new sites where mining can profitably be started.

Reverberatory furnace One where the flames pass above the matter to be melted, so that the heat beats down from the roof of the furnace.

Shaft mining The digging out of a vertical shaft, with galleries radiating out at each level where there is a mineral worth extracting. Some mines could provide more than one mineral.

Slate A hard rock that can be split into thin sheets, normally dark blue or grey in colour.

Slurry A mixture of rock fragments and much water. It is a convenient way of moving some rocks and clays as they can be pumped through pipe lines.

Smelting Separating metals from the waste rock or chemical compounds in which they occur as ores.

Truck payments The payment of wages either in goods or in company token coins that could only be spent in the company shop.

Trapper A child in charge of a door in a mine which controls the ventilation system.

Tonne 1 metric tonne = 0.984 of an imperial ton.

Undercutting Removing a slot at the base of a rock so that the weight of the mass above will bring it down.

Vein A seam of mineral; usually applied to non-ferrous metal ores.

Index

The numbers in **bold** refer to the figure numbers of the illustrations.